Network Slicing for 5G and Beyond Networks

S. M. Ahsan Kazmi • Latif U. Khan
Nguyen H. Tran • Choong Seon Hong

Network Slicing for 5G and Beyond Networks

S. M. Ahsan Kazmi
Institute of Information Security
and Cyber Physical Systems
Innopolis University
Innopolis, Russia

Department of Computer Science
and Engineering
Kyung Hee University
Seoul, South Korea

Nguyen H. Tran
School of Computer Science
The University of Sydney
Sydney, NSW, Australia

Latif U. Khan
Department of Computer Science
& Engineering
Kyung Hee University
Seoul, Korea (Republic of)

Choong Seon Hong
Department of Computer Science
& Engineering
Kyung Hee University
Seoul, Korea (Republic of)

ISBN 978-3-030-16172-9 ISBN 978-3-030-16170-5 (eBook)
https://doi.org/10.1007/978-3-030-16170-5

This Springer imprint is published by the registered company Springer Nature Switzerland AG.
The registered company address is: Gewerbestrasse 11, 6330 Cham, Switzerland

Preface

Network slicing will play an important role in enhancing the flexibility of the current and future cellular networks. In network slicing, the physical cellular network is divided into multiple virtual networks typically called a slice with heterogeneous capabilities. Each slice is then used to serve the end user. Network slicing gives us the ability to manage and adopt to the heterogeneous requirements imposed by the end users. Moreover, it is envisioned that, for the end users, 5G networks will incorporate many novel applications for vertical industries, each with its own set of stringent requirements. Thus, network slicing will play a pivotal role in meeting these set of requirements for each user (network applications) by allocation of a desired network slice that was not possible in legacy cellular networks.

This book will provide a comprehensive guide to the emerging field of network slicing and its importance to bring novel 5G applications into fruition. It discusses the current trends, novel enabling technologies, and current challenges imposed on the cellular networks. Resource management aspects of network slicing are discussed in great detail by summarizing and comparing recent network slicing solutions. Finally, it also presents a use case pertaining to offloading of data in augmented reality, which is an application for vertical industries. We hope this survey provides an overview on the subject and a direction for future research.

Innopolis, Tatarstan, Russia — S. M. Ahsan Kazmi
Seoul, South Korea — Latif U. Khan
Sydney, NSW, Australia — Nguyen H. Tran
Seoul, South Korea — Choong Seon Hong
January 2019

Acknowledgement

This work was supported by the National Research Foundation of Korea (NRF) grant funded by the Korea government (MSIT) (NRF-2017R1A2A2A05000995). Dr. CS Hong is the corresponding author.

Contents

Chapter 1
5G Networks

1.1 Introduction

Recent advances in wireless communication has enabled us in improving our tasks, financial aspects, well-being, and all the cutting edge enterprises. We assemble, dissect, and share information among various entities to have a profitable life which was impractical before the modern communication networks. Over the recent two decades, wireless communication has developed at an exceptionally quick pace to have a productive and completely digital future.

1.1.1 *Evolution of Cellular Systems*

The first generation (1G) of cellular systems was introduced in 1980 for voice transmission. Frequency division multiple access (FDMA) was used as access technique in the RAN of 1G systems. FDMA requires high gap between channels to avoid interference and each channel had the ability to serve only a single user. With the passage of time the number of users increased and 1G system suffered from the downsides of limited capacity, poor voice quality, and scalability issues. To overcome these downsides, 2G mobile technologies (i.e., GSM, D-AMPS) were developed that were based on time division multiple access (TDMA). Although TDMA allows multiple users per single channel using time sharing, still it requires to have large frequency gaps between the channels to avoid interference. Thus, to further enhance the performance of the cellular networks, third generation (3G) was introduced whose key features were higher capacity and data rates. The 3G uses the access technique of code division multiple access (CDMA) which is based on usage of the same frequencies among multiple users for simultaneous transmission. Although CDMA offers several advantages over its predecessor such as improvement in voice quality and better security, it required building of new

S. M. A. Kazmi et al., *Network Slicing for 5G and Beyond Networks*,
https://doi.org/10.1007/978-3-030-16170-5_1

infrastructure and expensive spectrum licensing fee for the operators. Moreover, the proliferation of novel access devices and introduction of new services such as video streaming and social media posed even higher QoS requirements for the operators, thus operators required to further enhance the capacity of the existing cellular systems in order to move from a voice driven system toward a data driven system. This led to the introduction of the fourth generation (4G) of cellular networks that are true IP-based system with even higher data rates and capacity. Moreover, many new features such as band aggregation, orthogonal frequency division multiplexing access (OFDMA) as an access scheme, multiple-input and multiple-output (MIMO) antennas, and beam forming are added in the 4G standard to enhance high data rates and capacity. Moreover, the number of connected devices with huge bandwidth requirements in cellular networks is expected to grow further due to technology adoption of industries and development of low cost Internet of Things devices.

According to *Mobile and wireless communications Enablers for Twenty-twenty (2020) Information Society* (METIS), there will be 33 times increase in the mobile traffic worldwide in 2020 compared to 2010 [1]. To cope with the gigantic increase in mobile data traffic as evident from the statistics, effective planning of the network is imperative. Furthermore, many efforts are being put by both the industry and academic worlds to meet the requirements posed by 5G networks [2]. Thus, the 5G standard will be driving the future cellular networks [3]. 5G cellular networks are intended to provide peak data rates of up to 10 Gbps, latency up to 1 ms, $1000\times$ number devices, 10 times energy efficiency, and high reliability, respectively, according to METIS.[1] As a result of massive technological revolutions, the demands posed by end users have increased drastically. To meet such drastic demands, ITU has classified the future 5G services into three main categories consisting of ultra-reliable low latency communication (URLLC), enhanced mobile broadband (eMBB), and massive machine type communication (mMTC) services. The existing mobile network architecture was designed to meet requirements for voice and conventional mobile broadband (MBB) services. Furthermore, the previous cellular generations were primarily designed to only fulfill the human communication requirements such as voice and data. However, 5G networks are expected to facilitate industrial communication as well in order to grow industry digitalization. Thus, enabling innovative services and networking capabilities for new industry stakeholders. The 5G technology is expected to provide connectivity and communication needs with specific solutions to vertical sectors such as automotive, health care, manufacturing, entertainment, and others in a cost-effective manner. Note that these novel services have very diverse requirements, thus, having traditional RAN and core management solutions for every service cannot guarantee the end user QoS. Moreover, some incremental improvement has been observed due to installation of small cells under macro cell coverage (i.e., heterogeneous networks (HetNets)) especially in congested locations. The concept of HetNets has already been implemented in current networks.[2] Other promising approaches for enabling

[1]https://bwn.ece.gatech.edu/5G_systems/.

[2]https://www.3gpp.org/technologies/keywords-acronyms/1576-hetnet.

the future RAN include enabling device to device (D2D) communication to reduce network traffic [4, 5], installing cache storage at the access networks to reduce delays, performing computation at the local base stations for real time analytics, and allowing to use unlicensed spectrum such as LTE-unlicensed (LTE-U) to further enhance the network capacity [6].

1.1.2 Heterogeneous Networks

As the number of devices in the cellular networks has been increasing at an alarming growth rate, a number of novel challenges pertaining to the mobile coverage, capacity limitations, and the required quality of service in the mobile networks need attention. To meet such challenges, a number of small cells are installed under the coverage of traditional macro cells. Such a setting is called the heterogeneous networks in which low power small-cell base stations (SBSs) are operated under the traditional macro cell coverage [7] as shown in Fig. 1.1. Many variants of SBSs, e.g., femto cells or pico cells can be adapted according to the network requirements in an area.

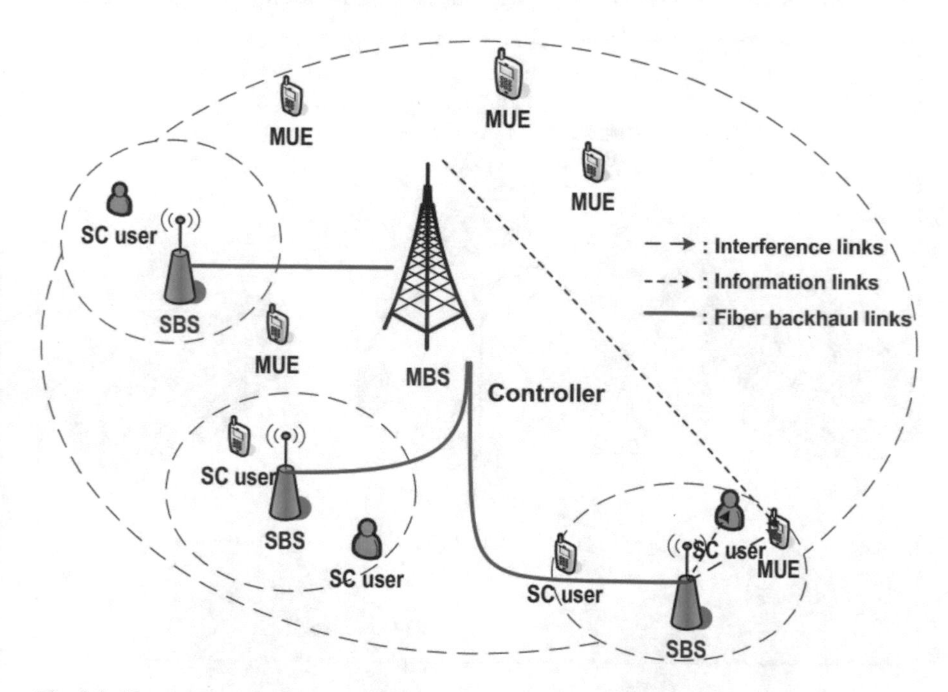

Fig. 1.1 Heterogeneous network consisting of MBS and multiple SBSs

HetNets contribute well in enhancing the spectrum efficiency, user coverage, and quality of service while saving the power and enhancing the spectrum resource efficiency. Moreover, HetNets have been widely adapted to provide better user experience in outdoor/indoor environments [8].

However, there exist a number of challenges such as interference management, resource allocation, and coverage area management among SBSs and the macro base stations. Various approaches have been proposed to meet these challenges including heuristic-based solutions, game theory-based solutions, matching theory-based solutions, and optimization theory-based solutions.

1.1.3 Device-to-Device Communication

Device-to-device (D2D) communication is one of the key technologies of 5G networks proposed to improve the spectral efficiency. In addition to the spectral efficiency, D2D has performed well in improving the user data rate, energy efficiency, and communication latency. D2D communication is base station independent, direct communication among two or more cellular users (Fig. 1.2). In other words, a D2D pair can directly communicate without routing its traffic through the central

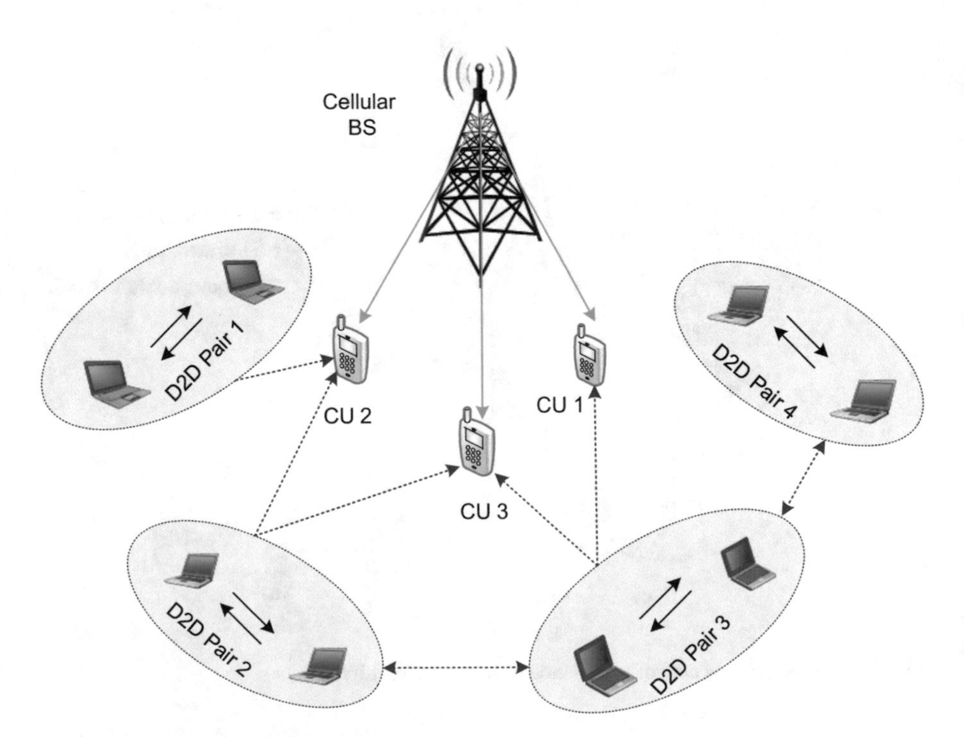

Fig. 1.2 A downlink D2D communication network

base station. D2D has found its applications mainly in proximity services, disaster situations, and local connectivity. Moreover, a number of study also exist for enabling D2D communication by utilizing the unlicensed spectrum.

Along with such benefits of D2D, there exist the challenges of interference management, resource allocation, and security in D2D and HetNets coexistence networks. As D2D communication is overlay on the existing cellular communication, the interference management and efficient resource allocation for these coexisted networks is challenging. Specifically, the aim is resource allocation in the coexisted networks by efficient coordination while minimizing the interference.

To meet such challenges, many centralized and distributed approaches have been proposed. Such approaches apply the interference avoidance schemes, conflict graphs to reduce interference, and game theory for the contention to access the licensed spectrum. Specifically, the resource allocation schemes are applied while mitigating the interference between D2D communication and regular cellular communication. The aim of such schemes is to maximize the overall utility of coexisted D2D and heterogeneous network.

1.1.4 LTE-Unlicensed

In order to meet the increasing data rate demands of 5G mobile users, LTE-unlicensed (LTE-U) is proposed where the LTE licensed spectrum is augmented with the unlicensed band. Specifically, LTE is enabled to operate on the licensed and unlicensed spectrum to meet the 5G data rate demands. It is obvious that such operation causes severe performance degradations in the pertaining unlicensed technologies, e.g., Wi-Fi (Fig. 1.3). Therefore a fair mechanism is required for the coexistence of LTE-U and Wi-Fi networks. Licensed assisted access has been standardized by 3GPP where LTE-U is enabled for downlink communication while the uplink communication and control signaling is performed on the licensed band.

Wi-Fi network is based on contention-based MAC known as CSMA where the network users contend for the channel access. Therefore, the random collisions degrade the spectrum efficiency. On the other hand, LTE-U is based on centralized MAC where the base station allocates the sub-channels to each cellular user. Therefore, the centralized MAC of LTE-U makes it more spectrally efficient as compared to the distributed MAC of Wi-Fi system which is the motivation of enabling LTE-U.

In order to provide sufficient fairness to the pertaining Wi-Fi networks, LTE-U is operated using channel selection or frame scheduling. In the channel selection, listen before talk (LBT) is used to find the cleanest channel for LTE-U operation. On the other hand frame scheduling adaptively varies the duty-cycle allocation among Wi-Fi and LTE-U to maintain Wi-Fi throughput up to some threshold.

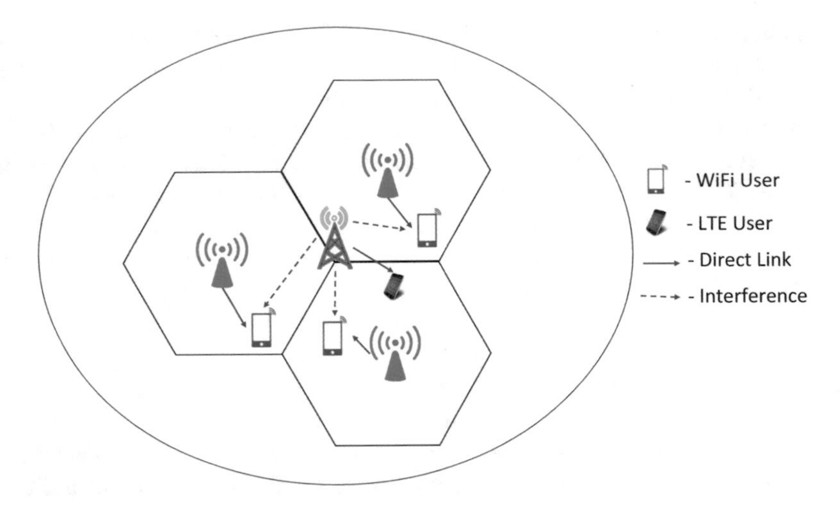

Fig. 1.3 LTE-U and Wi-Fi coexistence model

1.1.5 Non-orthogonal Multiple Access (NOMA)

In contrast to the orthogonal multiple access (OMA) scheme in which orthogonal channels are allocated to the cellular users in the same vicinity, non-orthogonal multiple access (NOMA) is proposed to pack multiple users on the same channel by utilizing the successive interference cancellation and efficient power control. NOMA brings the benefits of better connectivity and spectral efficiency. Moreover the limitations of connectivity in OMA scheme are alleviated by proposing NOMA-based communication.

NOMA utilized the successive interference cancellation by exploiting the channel gain differences of the cellular users in a network. In essence, users with difference channel gains are packed on a single channel by allowing them to transmit on different power levels (Fig. 1.4). In order to decode the signal successfully, successive interference cancellation is user after receiving the superimposed signal.

However, operating multiple users on a single channel causes severe interference issues. Efficient power control and user clustering is required to mitigate such interference issues. Efficient user clustering to exploit the channel gain difference can perform well in NOMA scheme. For that purpose, various machine learning tools for clustering are utilized in the literature. After the clustering, the efficient power control algorithms are applied for to achieve optimal cellular rates.

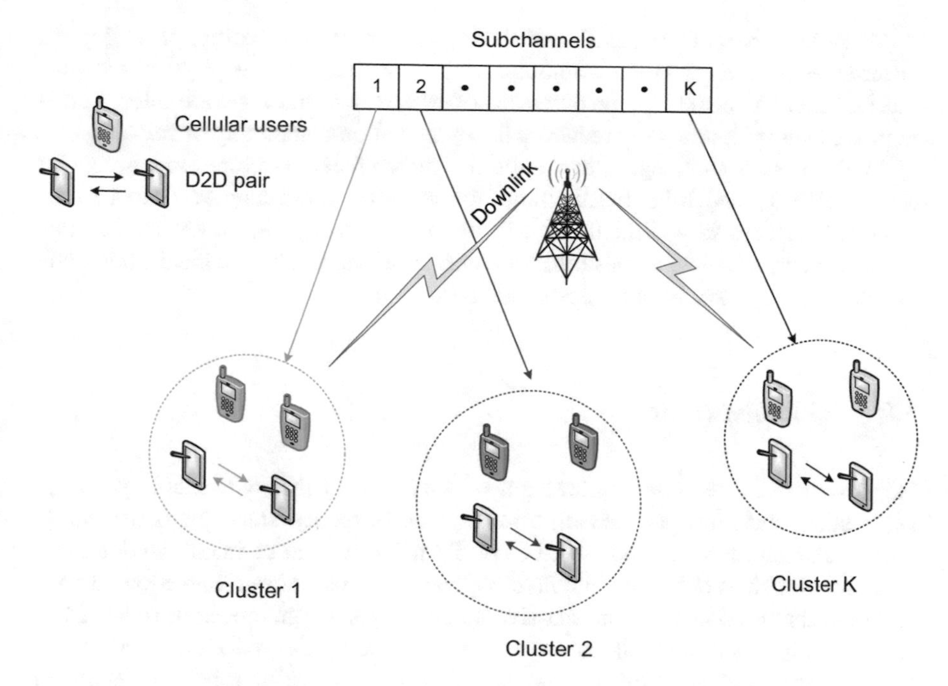

Fig. 1.4 System model of NOMA network

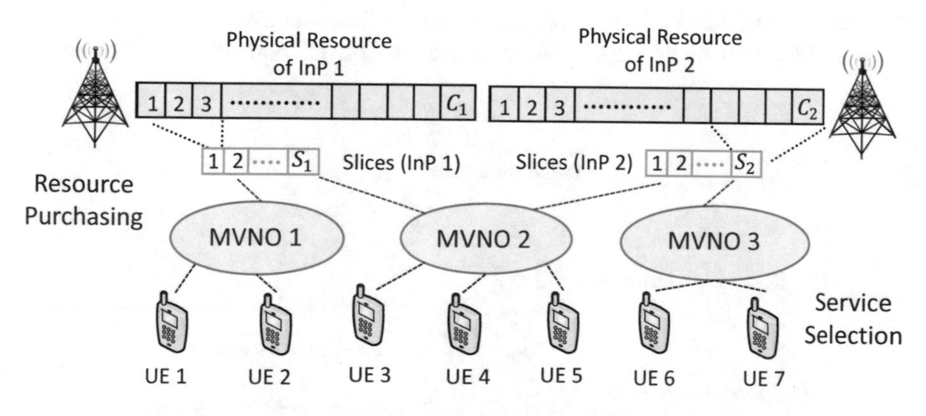

Fig. 1.5 System model for the allocation of virtualized physical resources to MVNOs

1.1.6 Wireless Network Virtualization

Wireless network virtualization (WNV) is a renowned 5G enabling technology which is proposed for efficient physical resource allocation. The aim of WVN is virtualization of physical resource of infrastructure providers (InPs) and efficient allocation of these virtualized resources to the mobile virtual network operations (MVNOs) (Fig. 1.5).

WNV has brought the benefits of flexibility in resource sharing, providing the differentiated services to the communication networks, and the physical resource abstractions. Moreover, the involvement of MVNOs in the resource allocation is improved while significantly reducing the computation complexity of InPs.

In WNV, it is challenging to handle the hierarchical structure among cellular users, MVNOs, and InPs. Furthermore, the resource purchasing according to the provided services is a difficult task to be performed by the MVNOs. For that purpose, many solutions consisting of auction-based games, hierarchical games, and matching theory have been proposed in the literature.

1.2 5G Deliverable

5G is envisioned as a novel paradigm of smart world that will enable powerful applications including self-driving cars, augmented reality, smart industries, smart homes, and smart cities. The use cases of 5G networks can be mainly divided into three types such as enhanced mobile broadband (eMBB), ultra-reliable low latency communication (URLLC), and massive machine type communication (mMTC) as shown in Fig. 1.6 [9]. Mobile broadband has use cases related to human access of data, services, and multimedia content. With the passage of time, the increased demand of users for mobile broadband tended to introduce enhanced mobile broadband (eMMB) that has more advanced applications than mobile broadband [10]. eMBB will enable applications (such as hotspot and wide-area coverage

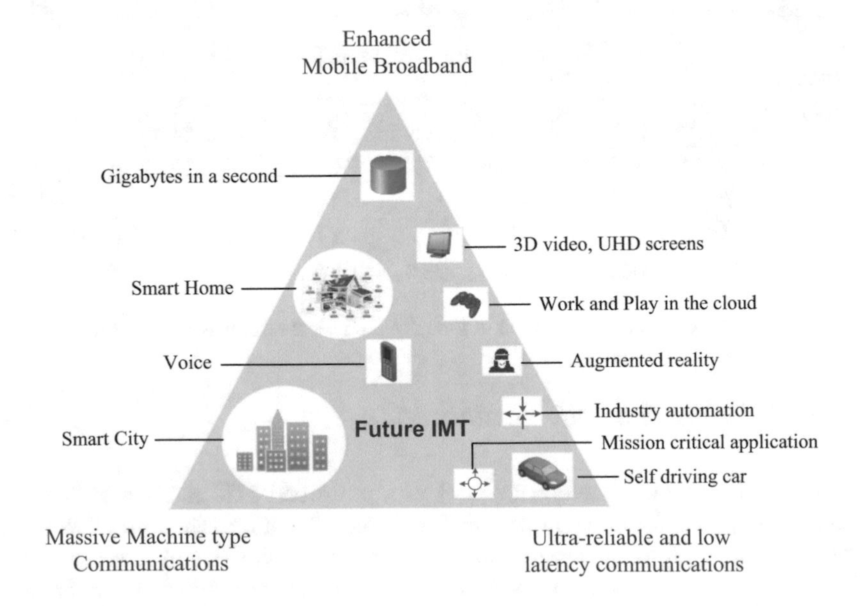

Fig. 1.6 5G use cases [9]

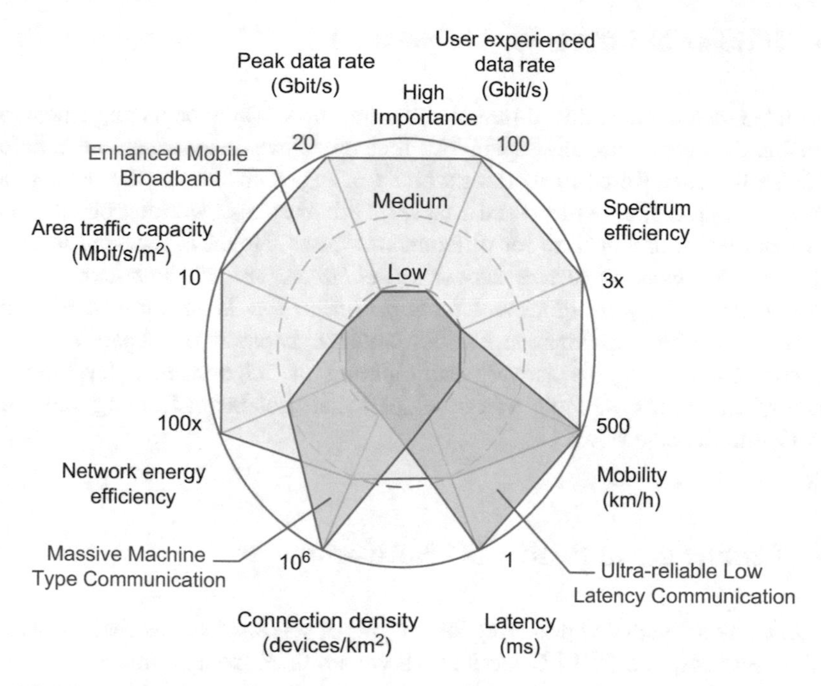

Fig. 1.7 Maximum 5G capabilities for different use cases [9]

scenario) with higher capacity, enhanced connectivity, and higher user mobility. Consider the hotspot scenario of a sports event, which consist of large number of users. To serve these users, requirements of low mobility and higher traffic capacity are needed. On the other hand, if we considered a bus, then there are requirements of high mobility and lower capacity than that of the hotspot. Apart from eMBB, mMTC is characterized by massive number of devices that transmit data with low volume and less sensitivity to delay. Other than mMTC, URLLC has strict requirements of higher throughput and low latency. Examples of URLLC applications include remote medical surgery and self-driving cars.

Consider Fig. 1.7, which illustrates the maximum values of specifications for different parameters such as peak data rates, spectrum efficiency, mobility, latency, connection density, network energy efficiency, and area traffic capacity [9]. Apart from the specifications for 5G, the proportion of capabilities compared to peak values for different use cases is also given in Fig. 1.7. Each capability might have different importance in different use case. For example, consider URRLC use case, the importance of mobility and latency is highest compared to other capabilities. On the other hand, mMTC has highest value for connection density than other capabilities. The notion of highest connection density is to ensure the connectivity of massive number of devices that might transmit low rate communication occasionally.

1.3 Industrial Efforts for 5G Networks

To turn 5G vision into reality, different industries are working on its implementation to enable the users with new capabilities that they have never experienced before. KT & SKTelecom, Korea are working with Samsung electronics on the development of 5G network. Nokia, Finland and Huawei, China are also working on enhancing the capacities of architecture for different use cases. Nokia is focusing to enable users with *5G mobility service supporting enhanced mobile broadband* and *5G mobility service supporting ultra-reliable and ultra-low latency communications*. Huawei is working to improve the 5G antenna capabilities. Apart from that, Ericsson, Sweden is also actively participating in 5G networks developments. Recently, Ericsson's 5G radio was developed that uses beam forming, multi-user MIMO, and massive MIMO.

1.4 Challenges to Realize 5G Networks

5G networks are aimed at providing three types of services such as eMBB services, mMTC services, and URLLC services. However, there exist certain challenges in its implementation that must be addressed.

1.4.1 Scalability and Reliability

Internet of Things (IoT) intends to seamlessly connect the world using smart devices that are heterogeneous in nature. In 5G, the IoT vision is supported more by technologies for mMTC use case applications implementation. To support massive number of IoT devices by 5G, it's imperative to design a scalable network architecture which is a challenging task [11]. Along with scalability, the design should also provide reliable operation.

1.4.2 Interoperability

In 5G networks, there will be a lot of heterogeneous IoT devices running different protocols. The major challenge in 5G is how to enable seamless interoperability between devices based on different technologies [12]. Enabling interoperability will perhaps allow users to actually realize IoT.

1.4.3 Sustainability

In 5G, sustainable network operation is imperative as mentioned earlier in this chapter to reduce the overall operation cost. Sustainability in 5G can be achieved either using energy efficient design or renewable energy sources. Along with this, 5G networks can also leverage energy harvesting to reduce their operation cost. However, designing energy efficient system with energy harvesting is a challenging task.

1.4.4 Network Slicing

5G networks will offer a novel applications with diverse requirements as explained in Sect. 1.2. Therefore, the already available mobile networks infrastructure would not be sufficient, which motivated the design of new architecture. To enable 5G applications with diverse requirements, network slicing can be utilized [13]. Network slicing enabled by network function virtualization and software defined networking is aimed at creation of slices of different types for diverse applications. One way to create slices is to assign a part of network end-to-end physical infrastructure resources to only one type of slices; however, his approach has disadvantage of supporting only one slices type by the physical infrastructure. Network slicing should allow the usage of same physical network infrastructure resources to enable the creation of different slices for different services. How to enable the creation of slices for different diverse applications using the same physical infrastructure is a challenging task.

1.4.5 Security

5G networks intend to rapidly rise the vertical industries, all of which impose wide variety of security challenges. For example, the distributed nature of self-driving cars makes them more susceptible to automotive cyberattacks. Other than that, the key enablers of 5G technology, NFV, and SDN will also add new security challenges. Apart from that, cloud computing and edge computing which are also the key enablers of 5G have security challenges. The centralized architecture of cloud computing has made it more susceptible to attack. Therefore, it's imperative to design new security mechanisms for 5G networks.

References

1. METIS. *Mobile and wireless communications system for 2020 and beyond (5G)*. Available: https://www.metis2020.com/documents/presentations/index.html
2. Chih-Lin, I., Rowell, C., Han, S., Xu, Z., Li, G., & Pan, Z. (2014). Toward green and soft: A 5G perspective. *IEEE Communications Magazine, 52*(2), 66–73.
3. Andrews, J. G., Buzzi, S., Choi, W., Hanly, S. V., Lozano, A., Soong, A. C., et al. (2014). What will 5G be? *IEEE Journal on Selected Areas in Communications, 32*(6), 1065–1082.
4. Kazmi, S. A., Tran, N. H., Ho, T. M., Lee, D. K., & Hong, C. S. (2016). Decentralized spectrum allocation in D2D underlying cellular networks. In *2016 18th Asia-Pacific Network Operations and Management Symposium (APNOMS)* (pp. 1–6). Piscataway: IEEE.
5. Kazmi, S. A., Tran, N. H., Saad, W., Han, Z., Ho, T. M., Oo, T. Z., et al. (2017). Mode selection and resource allocation in device-to-device communications: A matching game approach. *IEEE Transactions on Mobile Computing, 16*(11), 3126–3141.
6. Manzoor, A., Tran, N. H., Saad, W., Ahsan Kazmi, S. M., Pandey, S. R., & Hong, C. S. (2019). Ruin theory for dynamic spectrum allocation in LTE-U networks. *IEEE Communications Letters, 23*(2), 366–369.
7. Ho, T. M., Tran, N. H., Le, L. B., Saad, W., Kazmi, S. M. A., & Hong, C. S. (2016, May). Coordinated resource partitioning and data offloading in wireless heterogeneous networks. *IEEE Communications Letters, 20*(5), 974–977.
8. LeAnh, T., Tran, N. H., Kazmi, S. M. A., Oo, T. Z., & Hong, C. S. (2015, January). Joint pricing and power allocation for uplink macrocell and femtocell cooperation. In *2015 International Conference on Information Networking (ICOIN)* (pp. 171–176).
9. M Series. (2015). *IMT vision–Framework and overall objectives of the future development of IMT for 2020 and beyond*. Recommendation ITU-R M.2083-0.
10. Abedin, S. F., Alam, M. G. R., Kazmi, S. M. A., Tran, N. H., Niyato, D., & Hong, C. S. (2018). Resource allocation for ultra-reliable and enhanced mobile broadband IoT applications in Fog network. *IEEE Transactions on Communications, 67*(1), 489–502.
11. Alawe, I., Hadjadj-Aoul, Y., Ksentini, A., Bertin, P., & Darche, D. (2018, January). On the scalability of 5G core network: The AMF case. In *2018 15th IEEE Annual Consumer Communications Networking Conference (CCNC)* (pp. 1–6).
12. Konduru, V. R., & Bharamagoudra, M. R. (2017, August). Challenges and solutions of interoperability on IoT: How far have we come in resolving the IoT interoperability issues. In *2017 International Conference on Smart Technologies for Smart Nation (SmartTechCon)* (pp. 572–576).
13. Zhang, H., Liu, N., Chu, X., Long, K., Aghvami, A.-H., & Leung, V. C. (2017). Network slicing based 5G and future mobile networks: Mobility, resource management, and challenges. *IEEE Communications Magazine, 55*(8), 138–145.

Chapter 2
Network Slicing: The Concept

5G networks are intended to provide mainly three types of services: massive machine type communications (mMTC), ultra-reliable low latency communications (URLLC), and enhanced mobile broadband (eMBB). The mMTC is characterized by massive number of devices communicating with each other and requires low cost along with long battery backup time. Apart from that, URLLC need simultaneous low latency and ultra-reliability. On the other hand, eMBB requires higher data rates along with large coverage area. To enable users with eMBB, mMTC, and URLLC, it's imperative to redesign the network architecture. Network slicing is a promising candidate to allow 5G networks to offer a wide variety of services and applications such as e-health, augmented reality, smart transportation system, smart banking, smart farming, and mobile gaming. The notion of the network slicing is to leverage the network infrastructure resources for creation of multiple subnetworks for different types of services and applications. Then, each subnetwork will perform slicing of the physical network resources to yield an independent network for its applications. To enable different 5G services, slices of different types for mMTC, URRLC, and eMBB services that utilize the network resources can be created. We can assign each slice type with complete end-to-end network resources; however, it does not seem practical because of the high expense. On the other hand, it would be feasible to allow sharing of the network resources among multiple types of slices using technologies like network function virtualization and software defined networking.

2.1 Network Slicing: Concept and Definitions

The notion of network slicing is the virtual network architecture that will enable powerful and flexible capability of creating multiple logical networks on top of the common physical infrastructure as illustrated in Fig. 2.1. Network softwarization

S. M. A. Kazmi et al., *Network Slicing for 5G and Beyond Networks*,
https://doi.org/10.1007/978-3-030-16170-5_2

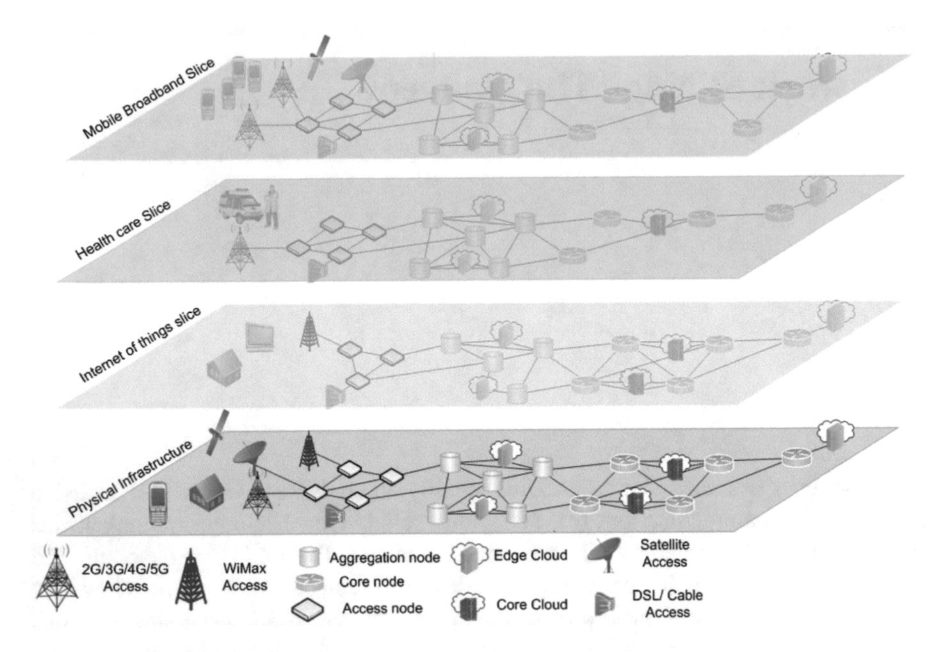

Fig. 2.1 5G network slicing [1]

is an emerging concept that will enable the network slicing using software-based solutions. Network softwarization can be achieved through technologies such as network function virtualization and software defined networking. Specifically, network slicing in 5G will use software defined networking, network function virtualization, cloud computing, and edge computing to enable flexible operation for different types of services over the same physical infrastructure. Network slicing allows the creation of logical networks for different types of services as illustrated in Fig. 2.1 [1]. Each logical network will have independent control with the ability of creation on demand.

The architectural vision of 5G infrastructure by public private partnership project (5G-PPP) proposed the division of network slicing architecture into five layers such as service layer, infrastructure layer, orchestration layer, business function layer, and network function layer [2]. On the other hand, the next generation mobile network (NGMN) alliance's architectural vision proposed the division of network slicing architecture into three layers: business application, business enablement, and infrastructure resource [3]. In [4], a generic framework for 5G network slicing is presented as shown in Fig. 2.2. This framework has three layers such as service layer, network function layer, and infrastructure layer. Apart from these three layers, it contains management and orchestration (MANO) entity. The MANO performs the translation of service models and use cases into slices. The network infrastructure layer deals with the physical network infrastructure that covers both core network and radio access network. Other than this, the infrastructure layer also performs infrastructure control and allocation of resources to the slices. Apart from

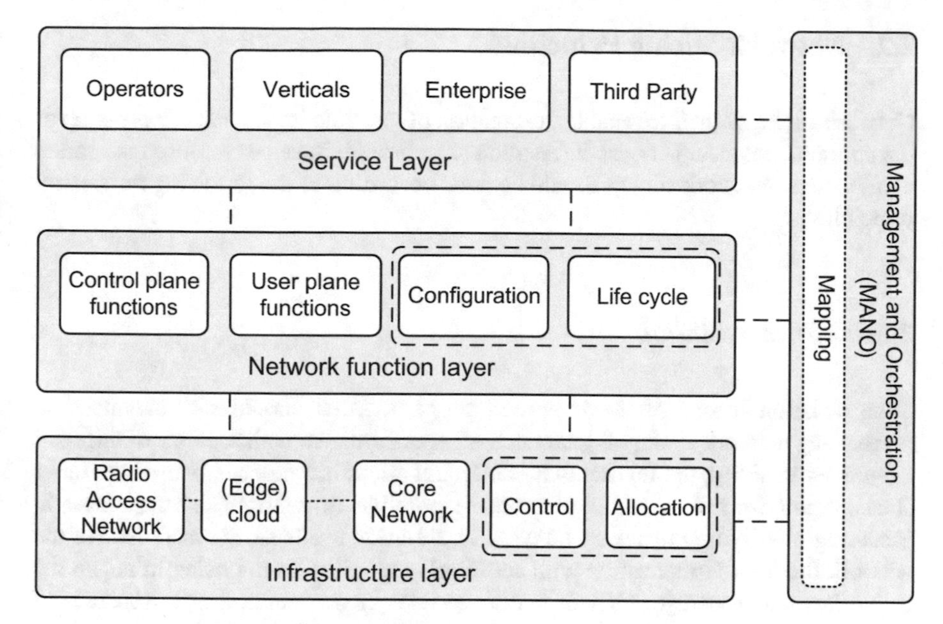

Fig. 2.2 5G network slicing layered framework [4]

infrastructure layer, the network function layer performs the encapsulation of the operations required in network functions configuration and life cycle management. These network functions are then chained together to deliver an end-to-end service. Following are the definitions of some terms that will be used later in this chapter.

Software Defined Networking Software defined networking allows the separation of control functionalities from the data transmission network. It actually divides the network into two planes such as data plane for data transmission and centralized control plane for performing control layer functionalities. The main advantage of software defined networking is the easier management due to its centralized nature; however, the centralized controller might suffer from the inefficiency when the scale of the network increases [5].

Network Function Virtualization Network function virtualization is envisioned to yield the architecture that allows virtualization of the functionalities of the network nodes, which are then chained together for performing different communication services. A virtual network function might comprise of one or more virtual machines running on different devices such as servers, edge computing nodes, cloud computing infrastructure, and network switches.

Tenant Tenant denotes the users that can access the shared resources with specific privileges and access rights.

Infrastructure Providers It refers to operators that provide the physical infrastructure and are responsible for maintenance and operation of the network.

2.2 Network Slicing Principles

Network slicing principles enable the creation of multiple logical networks on top of the common physical network infrastructure to provide services to different market applications. Network slicing should be done according to the following three main principles.

2.2.1 *Slice Isolation*

Slice isolation is one of the important properties that should be guaranteed in performing network slicing. It guarantees the performance with security of different tenants by enabling the feature that one tenant slices are independent of the other slices. Apart from that, isolation makes it possible to restrict the tenant user in accessing/modifying the other tenant slices. Isolation is a basic characteristic of the network function virtualization with additional capability of imposing limits on the network resources usage [6]. This feature enables guaranteeing the performance of different tenants by fair sharing of the resources.

2.2.2 *Elasticity*

Elasticity allows the dynamic alteration of resources allocated to different tenants slices in order to effectively utilize the resources. Elasticity can be realized by relocating the virtual network functions, scaling up/down of the allocated resources, and reprogramming the control and data elements functionalities [7]. The main challenge that will exist in the implementation of the elasticity is the policy for inter-slice negotiation so that the performance of different slices remains unaffected.

2.2.3 *End-to-End Customization*

Customization makes sure that allocated shared resources to different tenants are utilized effectively. It is expected that customization will be offered in future by many vertical industries [7]. To enable customization, network function virtualization, software defined networking, and network orchestration can be exploited for providing resource allocation in an agile way.

2.3 Network Slicing Enablers

Network slicing in 5G will be enabled by software defined networking, network function virtualization, cloud computing, and edge computing. 5G architecture is envisioned to provide slices of different services such as eMBB services, URLLC services, and mMTC services. This section discusses the key enablers of the network slicing.

2.3.1 *Software Defined Networking*

Traditional networking devices were designed to perform a set of functionalities and vendors need to update the design to enable new features. One solution is to build a custom hardware; however, this approach has disadvantage of high cost associated with building of custom hardware. Other than cost, significant efforts will be needed to come up with design that meet the most updated specifications. Once the system is designed, there is no way to make changes in it. All of the above factors motivated researchers to develop novel architecture that will enable configurable feature. SDN is a solution that is envisioned to enable users with separation of control plane from the underlying network devices, more granular security, lower operating costs, and reduced capital expenditure.

The scale of the network evolution imposes challenges on its management and configuration in a traditional network. SDN centralized architecture enables easier management of the large-scale networks [8] by decoupling the control plane from data plane as shown in Fig. 2.3. The logical control plane has the ability of easier configuration along with effective management functions. In software defined networking, the job of forwarding packets is performed by the centralized network controller through programmable interfaces. Floodlight, Onix, and NOX are the typically used SDN controllers [9]. OpenFlow is the most commonly used interface for controlling the packet flow [10].

The rise in scale of the network imposes stress on the centralized control in SDN. This has limitation effect on the OpenFlow networks scalability [11, 12]. To minimize load on the centralized controller, it is preferable to design an architecture that enables the processing of frequent events near the switches. In [13], a hierarchical architecture is proposed to reduce the load on the centralized controller. The controllers are divided into two categories such as local and root controllers. The local controller is connected to one or more switches, which is further controlled by the root controller. The root controller other than controlling the local controller performs functions that need the network wide view of the network. Apart from kandoo, Hyperflow [12] and Onix [14] also performed the distributed implementation of SDN to reduce load on the centralized controller.

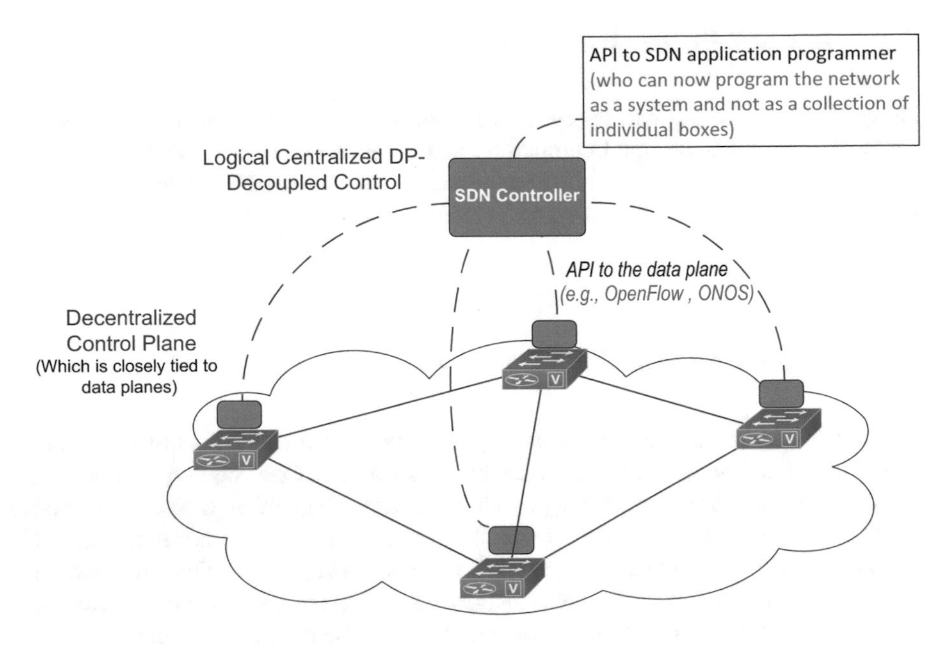

Fig. 2.3 Software defined networking

2.3.2 Network Function Virtualization

Network function virtualization (NFV) is a concept that leverages virtualization feature for transforming the network nodes functions into virtual functions that can be further chained together to enable different communication services. One or more virtual machines that run on different network nodes such as network switches and servers can be used to run a virtual network function. NFV uses commodity hardware to run software virtualization techniques for implementation of network functions as shown in Fig. 2.4. The virtual appliances have the advantage of instantiating without installation of new hardware. NFV will enable operators with the following advantages:

- Reduction of the capital expenditure by enabling operators with network functions without buying the hardware and payment as per usage.
- It also allows the operators to reduce the operational expenditure required for space, cooling, and power.
- It adds more flexibility to the system by enabling easier and quick scaling of the services.

In the context of network slicing, the framework of network function virtualization enables management of virtual network functions, service chaining [16], and latency oriented virtual network function embedding [7]. The architecture of network function virtualization given in [17] consists of three main components:

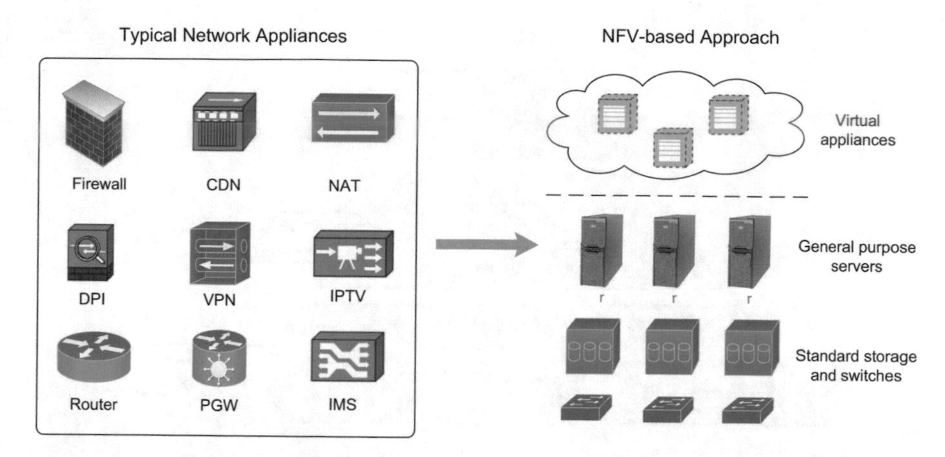

Fig. 2.4 NFV-based approach of instantiating network functions [15]

network function virtualization orchestrator, virtual network function manager, and virtualized infrastructure manager. The virtual network function manager performs the life cycle management of instances of virtual network functions. The virtual infrastructure manager is responsible for management and control of resources of network function virtualization infrastructure of network operator. The network function virtualization orchestrator performs orchestration of the network resources, validation and authorization of the virtual network function manager requests for resource of network function virtualization infrastructure, and life cycle management of the network service.

2.3.3 Cloud Computing

Cloud computing often called as cloud enables users with on-demand online pool of shared resources that include computing and storage resources, respectively. The cloud follows three types of delivery models such as infrastructure as a service (IAAS), platform as service (PAAS), and software as service (SAAS). SAAS allows users with on-demand use of softwares without installing it on a local personal computer. The software runs on cloud that is available for users and the users only pay for the software as per the use in a cheap way. Further, the users can access the software regardless of the location and enjoy more freedom of working in a group of peoples located at different places. The cloud software can be accessed either via client application or a browser. Examples of SAAS are dropbox, Gmail, Microsoft office 365, Facebook, etc. Apart from SAAS, PAAS allows the users with platform for development of applications and is made up of a database, browser, operating system, and programming language execution environment. PAAS enable the users to develop their applications regardless of taking care of

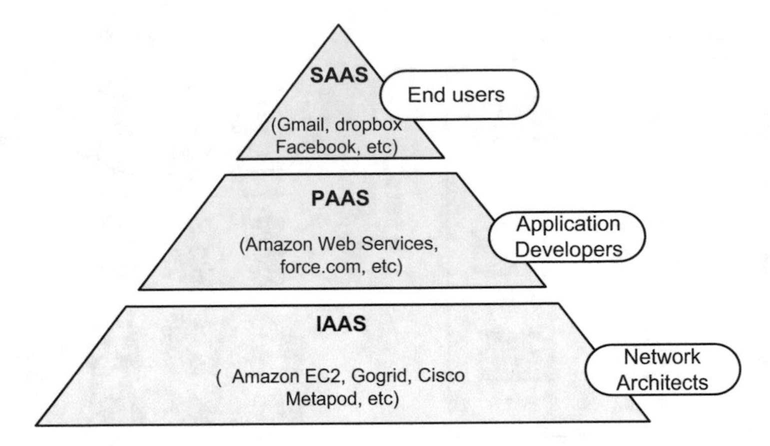

Fig. 2.5 Cloud models

the underlying infrastructure which is managed by the vendor. Examples of PAAS include Apprenda, windows Azure, Amazon web services, Engine Yard, AWS Elastic Beanstalk, and force.com. In IAAS, the virtualized computing resources are offered as a service via internet. Examples of IAAS include Amazon EC2, Gogrid, DigitalOcean, Cisco Metapod, and Google Compute Engine (Fig. 2.5).

Cloud Deployment Models

The cloud deployment models convey the nature and purpose of the cloud. Five types of cloud deployment models are identified by the NIST [18].

Private Cloud The computing and storage resources in a private cloud are owned by a single organization. In a private cloud, the users data is protected by the firewall and relationship between vendor and customer can be determined easily due to the reason of involvement of a single organization in the operation and management. Therefore, it is easy to detect the security risks. On the other hand, private cloud has disadvantage of operation, maintenance, and updating by a single organization, which might increases the overall cost.

Public Cloud The cloud services in a public cloud are owned by a third party service provider and provided through internet. All the software, hardware, and supporting infrastructure are managed and owned by the cloud service provider. This type of cloud deployment has the advantage of low cost and more reliability, but suffered from security concerns.

Community Cloud Community cloud allows several organizations to share the cloud infrastructure which is managed either by a third party or internally. The involved organizations worked for common business goals and similar needs. The use of community clouds is attractive for organizations that have constraint of regulatory compliance such as financial and e-health organizations.

Hybrid Cloud Hybrid clouds leverages the advantages of both private and public clouds to yield services with more flexibility. Private cloud can be used for services that require more security like processing of confidential information while public clouds can be used for services that need less security. Hybrid cloud enable users to leverage different cloud models which allow trade-off between different aspects of cloud computing. For example, a researcher will prefer to use private cloud for running critical workloads while prefer to use public cloud for database services.

Virtual Private Cloud Virtual private cloud is the shared pool of allocated resources in a public cloud with on-demand and configurable features. In virtual private cloud, the user data is isolated from other companies data using private IP addresses, cloud storage, and servers to achieve enhanced security.

Requirements for Cloud Computing in 5G

This sections explains the requirements for cloud computing architecture in 5G. These requirements can be divided into three categories such as provider requirements, enterprise requirements, and user requirements [19].

Reliability One of the most important requirements is the reliability that must be fulfilled in the cloud system. The cloud computing system might suffer from hardware and software flaws due to which it will not work properly. The system should be designed in way that must be resilient to both type of flaws because of the fact that some applications (such as vehicular communication and e-health) in 5G require ultra-high reliability.

Sustainability Sustainability deals with the use of renewable energy source and energy efficient hardware design. In 5G, we will witness a massive number of devices that will access the network, therefore, it's imperative to design the sustainable system that will fulfill the users requirements without adding pollution to the environment.

Scalability It deals with the management of increase in complexity due to an increase in computation tasks. Scalability in clouds can be either horizontal or vertical. Horizontal scalability is provided through load balancing. On the other hand, vertical scaling is to scale up capacity through increase in server resources. Most common way to do vertical scaling is to use expensive hardware and then utilize the hardware as a *virtual machine hypervisor*.

Fault Tolerance Fault tolerance is the ability of a cloud computing system to continue its operation if some of its component stops working. For example, Microsoft Azure encountered an outage of 22 h in March 13th and 14th, 2008 [19]. The cloud service provider should design the fault tolerant system that detect failure on time and then, apply mechanisms to recover from the outage within short time.

Security and Privacy Security and privacy should be considered in the design of the cloud computing system. Privacy defines the rules of who can access the users information while security is to protect the data from the attack of the malicious users. Cloud computing systems are more vulnerable to attacks due to its centralized nature, therefore, effective security algorithm should be designed to avoid the attack of the malicious user. A malicious attacker might attack the cloud and cause the outage of the service easily due to its centralized nature.

2.3.4 Edge Computing

Although cloud computing provides the advantages of scalability, elasticity, resource pooling, and sufficient storage capacity, the high latency associated with cloud makes it unattractive for real time and resource-intensive applications. To enable users with resource-intensive and latency sensitive applications such as augmented reality and e-health, edge computing comes into the picture which leverages pushing of the storage and computing resources to the edge of the network. In literature, three different terms such as cloudlets, fog computing, and mobile edge computing appeared till now and are based on pushing of the computing and storage resources to the network end. All of these terms represent edge computing with different context awareness, scope, and edge nodes nature features, respectively. The term cloudlet was coined by Satyanarayanan et al. to enable the placement of small-scale data center at the network edge for execution of resource-intensive applications [20]. Fog computing was coined by CISCO to refer to architecture that uses the edge devices (such as routers and gateways) to allow users with computing and storage capabilities [21]. The term mobile edge computing was coined by European Telecommunications Standards Institute (ETSI) to propose the placement of servers at the base stations [22].

2.4 Summary

5G networks are envisioned to offer three types of services such as mTC, URRLC, and eMBB. Each of these use cases has diverse network requirements. To enable these diverse use cases applications, network slicing is a promising candidate. Network slicing enables different applications by creating logical slices over the top of the physical infrastructure. This chapter describes the concept of network slicing. Moreover, slicing principles such as slice isolation, elasticity, and end-to-end customization are also explained. Finally, the key network slicing enablers such as software defined networking, network function virtualization, cloud computing, and edge computing are also explained.

References

1. Ordonez-Lucena, J., Ameigeiras, P., Lopez, D., Ramos-Munoz, J. J., Lorca, J., & Folgueira, J. (2017, May). Network slicing for 5G with SDN/NFV: Concepts, architectures, and challenges. *IEEE Communications Magazine, 55*(5), 80–87.
2. 5GPPP Architecture Working Group, et al. (2016, July). *View on 5G Architecture*. White Paper.
3. NGMN Alliance. (2015). *5G White Paper. Next Generation Mobile Networks*. White Paper (pp. 1–125).
4. Foukas, X., Patounas, G., Elmokashfi, A., & Marina, M. K. (2017, May). Network slicing in 5G: Survey and challenges. *IEEE Communications Magazine, 55*(5), 94–100.
5. Moon, S., LeAnh, T., Kazmi, S. M. A., Oo, T. Z., & Hong, C. S. (2015, August). SDN based optimal user association and resource allocation in heterogeneous cognitive networks. In *2015 17th Asia-Pacific Network Operations and Management Symposium (APNOMS)* (pp. 580–583).
6. Michalopoulos, D. S., Doll, M., Sciancalepore, V., Bega, D., Schneider, P., & Rost, P. (2017, October). Network slicing via function decomposition and flexible network design. In *2017 IEEE 28th Annual International Symposium on Personal, Indoor, and Mobile Radio Communications (PIMRC)* (pp. 1–6).
7. Afolabi, I., Taleb, T., Samdanis, K., Ksentini, A., & Flinck, H. (2018, third quarter). Network slicing and softwarization: A survey on principles, enabling technologies, and solutions. *IEEE Communications Surveys Tutorials, 20*(3), 2429–2453.
8. Hong, C. S., Ahsan Kazmi, S. M., Moon, S., Van Mui, N. (2016). SDN based wireless heterogeneous network management. In *AETA 2015: Recent Advances in Electrical Engineering and Related Sciences* (pp. 3–12). Cham: Springer.
9. Gude, N., Koponen, T., Pettit, J., Pfaff, B., Casado, M., McKeown, N., et al. (2008). NOX: towards an operating system for networks. *ACM SIGCOMM Computer Communication Review, 38*(3), 105–110.
10. McKeown, N., Anderson, T., Balakrishnan, H., Parulkar, G., Peterson, L., Rexford, J., et al. (2008). OpenFlow: Enabling innovation in campus networks. *ACM SIGCOMM Computer Communication Review, 38*(2), 69–74.
11. Curtis, A. R., Mogul, J. C., Tourrilhes, J., Yalagandula, P., Sharma, P., & Banerjee, S. (2011). DevoFlow: Scaling flow management for high-performance networks. *ACM SIGCOMM Computer Communication Review, 41*(4), 254–265.
12. Tootoonchian, A., & Ganjali, Y. (2010). HyperFlow: A distributed control plane for openflow. In *Proceedings of the 2010 Internet Network Management Conference on Research on Enterprise Networking* (p. 3-3).
13. Hassas Yeganeh, S., & Ganjali, Y. (2012). Kandoo: A framework for efficient and scalable offloading of control applications. In *Proceedings of the First Workshop on Hot Topics in Software Defined Networks* (pp. 19–24). New York: ACM.
14. Koponen, T., Casado, M., Gude, N., Stribling, J., Poutievski, L., Zhu, M., et al. (2010). Onix: A distributed control platform for large-scale production networks. In *OSDI* (Vol. 10, pp. 1–6).
15. Han, B., Gopalakrishnan, V., Ji, L., & Lee, S. (2015, February). Network function virtualization: Challenges and opportunities for innovations. *IEEE Communications Magazine, 53*(2), 90–97.
16. Lindquist, A. B., Seeber, R. R., & Comeau, L. W. (1966, December). A time-sharing system using an associative memory. *Proceedings of the IEEE, 54*(12), 1774–1779.
17. Quittek, J., Bauskar, P., BenMeriem, T., Bennett, A., Besson, M., & Et, A. (2014). *Network Functions Virtualisation (NFV)-Management and Orchestration*. ETSI NFV ISG, White Paper.
18. Singh, S., Jeong, Y.-S., & Park, J. H. (2016). A survey on cloud computing security: Issues, threats, and solutions. *Journal of Network and Computer Applications, 75*, 200–222.

19. Rimal, B. P., Jukan, A., Katsaros, D., & Goeleven, Y. (2011). Architectural requirements for cloud computing systems: An enterprise cloud approach. *Journal of Grid Computing, 9*(1), 3–26.
20. Satyanarayanan, M., Bahl, V., Caceres, R., & Davies, N. (2009). The case for VM-based cloudlets in mobile computing. *IEEE Pervasive Computing, 4*, 14–23.
21. *Fog Computing and the Internet of Things: Extend the Cloud to Where the Things Are*. (2015). Cisco White Paper.
22. Beck, M. T., Werner, M., Feld, S., & Schimper, S. (2014). Mobile edge computing: A taxonomy. In *Proceedings of the Sixth International Conference on Advances in Future Internet* (pp. 48–55). Citeseer.

Chapter 3
Resource Management for Network Slicing

3.1 Motivation and Introduction

Massive advancements in wireless networks have helped us in optimizing our operations, economics, health, and all modern industries. Among wireless networks, the cellular networks have received the most popularity [1].

A cellular communication network can broadly be categorized into two parts, the core network and the radio access network. This chapter focuses on the latter part of cellular communication, i.e., radio access network. A *radio access network* (RAN) is an integral part of the cellular communication system that implements the *radio access technology* (RAT) to enable the subscribers to access the *core network* (CN). A typical RAN consists of both wireless and wired links along with control or switching sites [2]. In a cellular system, the limited resources must be shared effectively among its subscriber to enhance the overall quality of service (QoS) of the network [3]. In the following subsection, we briefly present the evolution of cellular systems, shortcoming of previous generations, and 5G specifications along with the enabling technologies that will support in bringing the 5G networks into fruition. Then, we discuss about different types of resources available at the RAN of future cellular systems that require to be managed for effective operations.

3.1.1 Radio Access Network Resource Management

Resource management especially in the RANs is considered as the most crucial aspect of resource management due to limited and geographically distributed resources. Furthermore, the future RAN will be a very complex mixture of heterogeneous coexisting technologies that would require tight coordination for utilizing the shared and limited resources [4]. The resources in a 5G RAN can be broadly categorized into three categories: (1) spectrum and power resources

S. M. A. Kazmi et al., *Network Slicing for 5G and Beyond Networks*,
https://doi.org/10.1007/978-3-030-16170-5_3

available in radio access terminal for carrying out communication [5], (2) cache resources available for storage in the BS to reduce access delays, and (3) edge computing resources available at the access networks for performing real time computation.

The limited wireless spectrum in the RANs poses the biggest challenge that requires to be addressed. This limited wireless spectrum experiences severe congestion due the proliferation of new network devices and bandwidth hungry applications [6, 7]. To address this issue, new spectrum above 6 GHz is considered by 5G new radio (NR) for addressing specific use cases requiring extremely high data rates. Note that bands below 6 GHz are crucial to support wide-area coverage and 5G scenarios of mMTC. The 5G NR would be capable to operate on multiple bands to address diversified requirements from the envisioned 5G usage scenarios. Moreover, power management on this spectrum will also play a very crucial role for enhancing the performance. Therefore, the main goal is to develop efficient RAN schemes for spectrum and power allocation that can efficiently perform resource management [8].

Caching is used to store the popular contents temporarily in a cache to enable lower latency and reduced backhaul usage. According to statistics, video streaming traffic aids a significant portion of 54% to the whole INTERNET traffic which is intended to grow up to 71% in 2019 [9]. Most often the users request similar video contents frequently which induces overhead on the backhaul networks. Caching the content at the BS will reduce the congestion of the backhaul links and latency experienced by the user, respectively. However, cache space is very limited compared to the number of contents. This requires to design efficient caching schemes for saving popular contents that can be reused later with minimum delay. Thus, efficient caching schemes are required for 5G RANs that will play a vital role in bringing these novel services such as URLLC service into fruition.

The 5G network will require to perform a number of computation tasks especially after the advent of Internet of Things (IoTs). Efficient computation will play a vital role in realizing a number of real time services. One option is to adopt cloud computing for such services. Cloud computing enhances the user's overall QoE by providing the shared computing and storage resources online as a service in elastic, sustainable, and reliable manner. Although cloud computing offers significant advantages, latency sensitive applications such as self-driving cars, mission critical applications, industrial automation, and augmented reality incur performance degradation because of the distant locations of the cloud. Edge computing is a solution to delay sensitive application that pushes the computing resources to the edge of the network. In [10], Satyanarayanan et al. presented the concept of cloud-let, a small-scale data center positioned at the network edge to enable the execution of resource-intensive applications with low latency. In [11], CISCO coined the term fog computing, an architecture based on usage of edge devices to enable local computing resources. Mobile edge computing (MEC) was the term coined by European Telecommunications Standards Institute (ETSI) [12], to allow the placement of storage and computing resources at the base station (BS).

To fulfill these diverse heterogeneous requirements posed by vertical industries, a novel and promising concept of network slicing can be adopted. Network slicing enables one physical network into multiple, virtual, end-to-end (E2E) networks, each logically isolated including device, access, transport, and core network. A slice is dedicated for different types of service with characteristics and requirements pertaining to the required service by the vertical industry.

3.1.2 Network Slicing

Network slicing has recently attracted significant attention due to its wide applicability for 5G networks [13]. Through network slicing, the physical infrastructure of a cellular operator can be logically divided into slices of heterogeneous capabilities that can support various services with different requirements [14, 15]. For example, an operator can provide a dedicated slice of specific network capabilities for handling augmented reality applications with ultra-reliable and low latency communications and another dedicated slice with different network capabilities for video-on-demand services that require high throughput. This enables network operators to provide network-as-a service and enhances the network efficiency. One important aspect in enabling network slicing is to enforce strong isolation between different slices such that any actions in one slice does not affect other operating slices.

Network slicing is also being supported by 3GPP by defining a novel network architecture that supports slicing. In particular, the 3GPP working group SA2 has already defined the basis for building an evolved core network infrastructure managing multiple slices on the same network infrastructure [16]. Network slicing can be easily realized due to the revolutionary technologies of software defined networks (SDN) and network function virtualization (NFV). The work in [17] presents a detailed survey on wireless resource slicing and the challenges associated with isolations of wireless slices. Similarly, other notable works related to novel network slicing architectures can be found in [18–20]. All these aforementioned works basically discuss how to enable network slicing and its benefits in 5G networks. Specifically, they discuss the roles and interactions of network slicing enablers such as software defined networking (SDN) and network function virtualization (NFV) in the 5G network architecture. They do not consider specific algorithmic approaches for slice creation, slice allocation, slice interactions, admission control, etc. Moreover, considering the potential benefits of network slicing a number of mobile operators have shown keen interest and efforts in adopting network slicing in their network. SK Telecom and Ericsson successfully demonstrated network slice creation and operation for augmented reality solutions in 2015. Similarly, a proof of concept for 5G core was provided in 2016 by Ericsson and NTT DoCoMo by enabling dynamic network slicing. Other notable efforts pertaining to network slicing have been demonstrated by operators such as Huawei, Deutsche Telekom, ZTE, and China Mobile.

Even though there is technology development on network slicing, the majority of them focused on slicing at the core network due to massive advancements in SDN, virtualization, and NVF technologies. On the other hand, there are limited researches in the radio access network (RAN) slicing, in which most of them focused on slicing a single RAN resource, e.g., either spectrum or base station (BS) [18–22]. Note that network capacity is inarguably a critical resource of RAN, however, other resources of RAN such as cache space [23, 24], backhaul capacity, and computing at the RAN also need to be considered for RAN network slicing such as the work in [25] considers caching and backhaul limitation of 5G RANs.

3.2 RAN Resources

This section provides overview of the current advances in resource management of the RAN of the next generation cellular networks. In this work, we specifically discuss about three types of important resources available in the RAN, i.e., radio resources, cache resources, and MEC resources.

3.2.1 Radio Resources

The efficient spectrum management in 5G cellular networks is imperative to enable mobile subscribers with higher data rates, reduced latency, and ultra-reliability. The radio resource management has two main challenges of spectrum scarcity and spectrum resource allocation [26]. To tackle the issue of spectrum scarcity, researchers are trying to explore new high radio frequency bands. In 5G, cellular networks technology is intending to use millimeter wave (mmWave) frequencies band to enable higher data rates [27]. Other than spectrum scarcity, we can leverage optimal spectrum management to improve the quality of service [28]. Some recent proposals are discussed in the following paragraph.

In [29], Adedoyin et al. considered a two-tier network consisting of macro cells and dense deployed femto cells with co-channel assignments. They have formulated the problem of radio resource allocation as an optimization problem which is of mixed-integer nonlinear programming (MINLP) type and whose objective is to minimize the total transmit power. The MINLP problem is transformed into a simple problem as a mixed integer linear Programming using reformation-linearization technique and then, QoS-based radio resource management algorithm is proposed to minimize the interference and improve QoS. The key contribution of their work lies in the consideration of the joint resources allocation and power control for femto cells. The problem of resource allocation in two-tier HetNets is also considered in [30]. In resource allocation, the macro cell protection is achieved through the use of constraints on cross-tier interference. The problem of resource allocation is formulated as *mixed-integer resource allocation problem* which is solved using two different algorithms. One algorithm is based on duality-based optimization approach and the other one is proposed based on matching theory.

In [31], a downlink resource allocation in two-tier HetNets using the same frequency for small cells and macro cells is considered. The objective function is to minimize the total power that jointly takes into account the circuit power and transmission power. The proposed energy efficient scheme uses both progressive analysis and linear integer binary programming. Their proposed scheme allows the switching off the underutilized small cells which increase the energy efficiency. Liu et al. in [32] proposed an algorithm based on stable matching theory for resource allocation in two-tier HetNets. The problem is formulated to maximize the sum rate of the femto cell user equipment and then, matching game-based algorithm is proposed for resource allocation.

3.2.2 Caching

Caching is used to store the popular contents temporarily in a cache at intermediate locations in a network to jointly enable the fast access and elimination of duplicate transmission. Most often the users request same video contents frequently, which induce significant overhead on the backhaul networks. Caching the content at the BSs will reduce jointly the congestion of the backhaul links and latency experienced by the end user. The architecture of cache enabled cellular networks can be divided into four categories, i.e., *cache enabled C-RAN*, *cache enabled macro cellular networks*, *cache enabled D2D networks*, and *cache enabled HetNets*, respectively [33]. Other than position of cache installation, the control structure of the caching in cellular networks can be either centralized or decentralized. In a centralized structure, the placement and delivery of the content is controlled by a centralized entity while in decentralized control, the placement and delivery of the content is controlled by the BS or the user equipment. The choice of control structure depends on the trade-off between the complexity and achievement of global optimal solution. The decentralized algorithms have lower complexity than centralized one but might not achieve global optimal solution.

In 5G HetNets, cache can be placed at different positions in a radio access network such as at macro cell BS, small cell BS, and devices, respectively. In [34, 35], cache is placed at the BS in a single tier network. The authors in [34], proposed a caching scheme based on usage of BSs collaboration for minimizing the aggregated cost required to deliver files. They have considered both coded (content can be stored at multiple BSs) and non-coded caching (content can be stored only at a single BS if cached otherwise no caching is considered). The problem of caching for non-coded case is formulated as integer program to minimize the aggregated cost and is proven to be NP-complete. Then, a near-optimal algorithm using dynamic programming is proposed. On the other hand for the non-coded caching scenario, the problem is formulated as linear program which has a polynomial time solution. The limitation of their work lies in usage of information about the popularities of the contents whose accurate estimation might be difficult. In [35], energy efficient caching is proposed to jointly optimize the cache placement and cache hit ratio. A technique using short noise model (SNM) is proposed to estimate the content popularity. The cache hit rate is then improved by proposing distributed caching policy. Along with this, the authors have formulated caching problem as an optimization problem and presented a closed-form solution that maximizes the energy efficiency for the optimal cache capacity. In [34, 35], the authors considered the cache located at BSs of a single tier network. On the other hand, cache can be placed at both macro cells and small cells which allows the users to access the contents from the nearest BS. Apart from small cells in HetNets, cache enabled D2D communication enables the users to communicate without traversing the BS with each device having the capability of cache.

Yang et al. in [36] considered a three tier network that consists of D2D pairs, relays, and BSs. The node (uses, relays, and BSs) locations are first modeled as mutually independent Poisson point processes and then, protocol for content access is proposed. Along with this, the authors derived analytically expressions for outage probability and average ergodic rate for subscribers considering different cases. Further, derivation of delay and throughput using continuous-time Markov process and multiclass processor-sharing model are also performed. The authors considered unicast point-to-point transmission, which minimizes the backhaul traffic but fails to reduce "on air" congestion effectively. Along with this, the wireless medium broadcast nature in contrast to wired medium is not effectively exploited.

3.2.3 Edge Computing Servers

Edge computing is a solution to delay sensitive application that pushes the computing resources to edge of the network. In [10], Satyanarayanan et al. presented the concept of cloud-let, a small-scale data center positioned at the network edge to enable the execution of resource-intensive applications with low latency. In [11], CISCO coined the term fog computing, an architecture based on usage of edge devices to enable local computing resources. Mobile edge computing (MEC) was the term coined by European Telecommunications Standards Institute (ETSI) [12], to allow the placement of storage and computing resources at the base station (BS). A comprehensive survey of MEC is presented in [37]. Resource management in MEC can be divided into three categories such as single user MEC systems, multiuser MEC systems, and MEC systems with heterogeneous servers, respectively. Other than that, offloading can be either binary offloading that is based on offloading the whole task to the MEC server or partial offloading that is based on offloading the partial task to the MEC server for execution.

The partial task offloading has been considered in [38, 39]. You et al. in [39] studied the resource allocation for multiusers in orthogonal frequency division multiple access (OFDMA) and time division multiple access (TDMA) based multiuser MEC systems, respectively. The resource allocation problem for TDMA based MEC system is formulated as a convex optimization problem for minimizing the summation of the weighted user equipment energy considering the constraints of computation latency, MEC server computation capacity, and time sharing, respectively. The priority-based optimal resource allocation strategy is proposed that is based on assigning higher priority to users having insufficient local computation resources to run the applications within allowed time. On the other hand, the resource allocation problem for OFDMA-based multiuser MEC is formulated as a mixed-integer problem. This problem is solved through sub-optimal algorithm having low complexity which is based on transformation of OFDMA problem into its counterpart TDMA problem. The key advantage of the priority-based algorithm is consideration of both channel gain and local computing energy of multiple users; however, high complexity is associated with the proposed algorithms when the capacity of the MEC is considered finite, which seems more practical than considering MEC servers with infinite capacity to reduce the complexity of the proposed algorithm. In [38], the authors proposed a partial offloading system for a single user MEC system. A cost function is formulated for a single user application that incorporates both local execution time and offloaded execution time. The cost of the local execution only considers the local execution delay while the offloaded execution cost is weighted sum of three factors such as: delay occurred in offloading and sending the task back to the user, usage of computational resources at small cell MEC server, and radio resource usage. The objective function is to minimize the sum of the cost function for all components of a task. The authors proposed a deep supervised-based learning approach to solve the partial offloading problem. Their proposed approach has lower complexity such as $O(mn)^2$ for n task components of an application with m hidden layers, than the exhaustive algorithm which has complexity $O(2^n)$. The prime limitation of the proposed approach lies in requirement of large data set for training system prior to use.

The binary task offloading has been considered in [40, 41]. The authors in [40] considered a single user MEC system with another energy harvesting. The authors defined an execution cost function which is the weighted sum of the task dropping cost and execution delay. The dropping cost reflects to the tasks that are dropped either due to deep fading or due to lack of energy at the mobile device, which is only powered by the harvested energy in their system model. The problem is formulated as optimization problem to minimize the time average of the execution cost. Further, they have proposed online Lyapunov optimization-based approach to solve the optimization problem. The proposed online algorithm has advantage of non-usage of distribution of the computation task request process and channel statistics. The main disadvantage of the proposed system is usage of only harvesting source for powering the mobile device. There are significant random variations in the harvesting energy either from the natural sources or radio frequency sources. Therefore, it is practically more feasible to use hybrid energy sources that utilize both harvested energy and grid energy to enable the continuous operation of the devices. In [41], the fleet of drones (user's nodes) running resource-intensive tasks such as identification and classification of objects is considered. A cost function to jointly minimize the delay and energy consumption is defined and then, a game theory-based distributed offloading scheme is proposed. The theoretical game with three different strategies (such as local task computation, task offloading to a BS through a local wireless network, and offloading of task to a server through cellular network) and drones as players is considered to perform task offloading. The proposed approach has advantage of its distributed nature; however, its associated complexity has not been discussed in the paper.

3.3 Use Case: Virtual Reality (VR)

Virtual reality (VR) is an interactive digital real time experience taking place with a simulated artificial environment. It mainly includes audio and visual interactions. The goal of a VR system is to generate a real time virtual environment that mimics the human perceptions. To support this real time virtual environment, the wireless system needs to cope with massive amount of bandwidth and latency requirements which was not possible with the previous cellular generations. Indeed some VR technologies such as VR goggles are already emerging, however, the full potential of VR systems to achieve a fully immersive experience is yet to be explored. A detailed work on VR requirements, its enablers, and its challenges are discussed in [42]. In this work, we focus on one of the most important requirements of VR, i.e., offloading computation task. In this use case, we consider the computation offloading use case of a VR. Migrating computationally intensive tasks from VR devices to more resourceful cloud/fog servers is necessary to increase the computational capacity of VR devices.

3.3.1 System Model

Let $\mathcal{N} = \{1, \ldots, N\}$ be the set of N mobile users with VR capability as shown in Fig. 3.1, and $\mathcal{M} = \{1, \ldots, M\}$ be the set of M MEC servers. Each mobile user $i \in \mathcal{N}$ has a computational task to offload that is denoted by $\tau_i = \{S_i, T_i, C_i\}$, where T is the worst case execution time of the task, C is the budget of user i for offloading

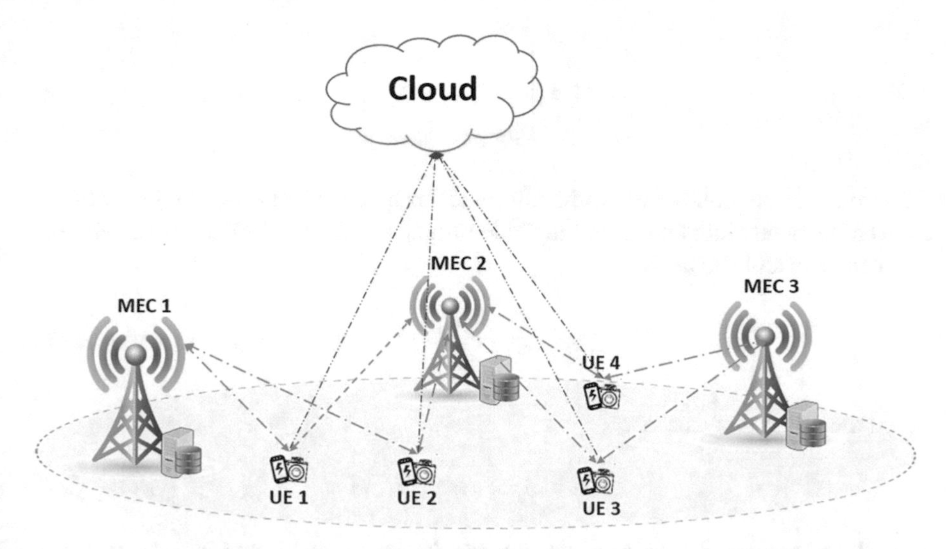

Fig. 3.1 System model

the task, and S_i is size of the task. Moreover, we assume the task to be a video task. Typically, in a video frame, there are four components: horizontal pixels, vertical pixels, frame rate, and color depth, therefore, we consider $S_i = \{H_i, V_i, F_i, L_i\}$, where H is horizontal pixels, V is vertical pixels, F is number of frames per second, and L is the length of video. Thus, the size of video is calculated using the following equation:

$$S_i = \frac{H \times V \times F \times L \times 3}{1024 \times 1024} \tag{3.1}$$

In this work, we assume that the bandwidth allocated for user i from different MEC j is different, $d_{i,j} \neq d_{i,k}, \forall i \in \mathcal{N}, \forall j,k \in \mathcal{M}, j \neq k$. This difference depends upon factors such as location, distance, and power level of devices. In our model, we consider two main factors pertaining to a user that affect the offloading decision: processing cost of the task and the transmission time of the task. Thus, we have two cases for any user $i \in \mathcal{N}$. The first case occurs when user i directly uploads to the cloud server. In this case, let c_{i0} be the cost for data storage where c_{i0} is calculated using a linear function with coefficient δ. The goal of assuming a linear cost function is to enable the pay as you go policy. This means the more data is uploaded, the more cost is required to be paid. The second case occurs when user i uploads the task to an MEC server. Note that each MEC server has a limit on its processing capacity which is denoted by Γ_j. Now, we define a bandwidth allocation matrix for the uplink transmission:

$$\mathbf{U} = \begin{bmatrix} u_{11} & \dots & u_{1M} \\ \vdots & \ddots & \vdots \\ u_{N1} & \dots & u_{NM} \end{bmatrix} \tag{3.2}$$

where u_{ij} is the uplink bandwidth allocated for user i when offloading to MEC j. Given the bandwidth and size of the video frames x_{ij}, the uplink transmission time t_{ij}^{u} can be calculated as follows:

$$t_{ij}^{u} = \frac{x_{ij}}{u_{ij}}. \tag{3.3}$$

And the processing cost as:

$$c_{ij} = \delta_j \times x_{ij}.\forall j \in \mathcal{M} \tag{3.4}$$

Similar to the cloud case, here also we use the linear cost function for calculating the processing cost in order to follow the pay-as-you-go policy and δ represents the coefficients of the cost function.

Then, the bandwidth allocation matrix for the downlink transmission can be represented by:

$$\mathbf{D} = \begin{bmatrix} d_{11} & \dots & d_{1M} \\ \vdots & \ddots & \vdots \\ d_{N1} & \dots & d_{NM} \end{bmatrix} \tag{3.5}$$

where u_{ij} is the downlink bandwidth allocated for user i when associated with MEC j. Given the size of output $S_i^{\prime}$ from MEC j, the downlink transmission time can be defined as follows:

$$t_{ji}^{d} = \frac{\eta x_{ij}}{d_{ij}}, \tag{3.6}$$

where η is the coefficient representing the relation between input and output data. For example, $\eta = 0{:}5$ mean that the result is equal to 50% of input data. Then, the total transmission cost is defined as follows:

$$t_{ij} = \left(t_{ij}^{u} + t_{ji}^{d}\right) \tag{3.7}$$

Note that the transmitted and received time for a user i from the central cloud server is denoted by t_{i0}^{u}, and t_{0i}^{d}, respectively. Moreover, we assume that the demand of users is to be served at the MEC for low processing time given it has enough capability. Therefore, to minimize the total cost of the end user, we normalize the total transmission and processing cost functions by dividing it with the original cost, i.e., cost required for a user i when it uploads its task to the cloud server.

$$t_i' = \left(\frac{\sum_{j \in M} t_{ij}^u}{t_{i0}^u} + \frac{\sum_{j \in M} t_{ji}^d}{t_{i0}^d} \right), \forall i \in N \tag{3.8}$$

and the processing cost

$$c_i' = \frac{\sum_{j \in M} c_{ij}}{c_{i0}} . \forall i \in N \tag{3.9}$$

3.3.2 Problem Formulation

In this subsection, we formulate our problem of task offloading for the VR enabled users. We aim to minimize the total transmission and processing cost of the network. Then, our problem can be stated as follows:

$$\underset{x}{minimize} : \sum_{i \in N} \alpha \left(\frac{\sum_{j \in M} t_{ij}^u}{t_{i0}^u} + \frac{\sum_{j \in M} t_{ji}^d}{t_{i0}^d} \right) + (1 - \alpha) \frac{\sum_{j \in M} c_{ij}}{c_{i0}} \tag{3.10}$$

$$subject\ to : \sum_{j \in M} x_{ij} = \frac{H_i \times V_i \times F_i \times L_i \times 3}{1024 \times 1024}, \forall i \in N, \tag{3.11}$$

$$\sum_{i \in N} x_{ij} \leq \Gamma_j, \forall j \in M, \tag{3.12}$$

$$\sum_{j \in M} t_i \leq T_i, \forall i \in N, \tag{3.13}$$

$$\sum_{j \in M} c_i \leq C_i, \forall i \in N, \tag{3.14}$$

$$t_i' < 1, \forall i \in N, \tag{3.15}$$

$$c_i' < 1, \forall i \in N, \tag{3.16}$$

$$x_{ij} \geq 0, \forall i \in N, \forall j \in M. \tag{3.17}$$

Our objective function (3.10) considers both the transmission and processing cost in which α represents the trade-off coefficient that ranges from [0, 1]. Constraint (3.11) guarantees that all demand is served by the network whereas the constraint (3.12) represents that the MEC capacity is not violated. Furthermore, the constraint (3.13) states that the budget of any user does not exceed the processing cost. Lastly, constraints (3.15), (3.16) state that the cost of transmission and processing at MEC are less compared to that of a central cloud, respectively. To solve this problem we use the alternating direction method of multipliers (ADMM) approach. ADMM has the capability of parallel solving the designed problem in a distributed fashion in which each user or MEC server will solve their individual variable(s). Next, we describe our designed approach.

3.3.3 ADMM-Based Solution

First we define the objective function (3.10) as follows which is equivalent to:

$$(3.10) = \alpha \left(\sum_{j \in M} \left(\frac{x_{ij}}{u_{ij} t^u_{i0}} + \frac{\eta x_{ij}}{d_{ij} t^d_{ji}} \right) \right) + (1-\alpha) \left(\frac{\sum_{j \in M} \delta_j x_{ij}}{c_{i0}} \right) \tag{3.18}$$

$$= \alpha \left(\sum_{j \in M} \left(\frac{1}{u_{ij} t^u_{i0}} + \frac{\eta}{d_{ij} t^d_{ji}} \right) \right) x_{ij} + (1-\alpha) \left(\frac{\sum_{j \in M} \delta_j}{c_{i0}} \right) x_{ij} \tag{3.19}$$

$$= \left(\alpha \sum_{j \in M} \left(\left(\frac{1}{u_{ij} t^u_{i0}} + \frac{\eta}{d_{ij} t^d_{ji}} \right) + (1-\alpha) \left(\frac{\sum_{j \in M} \delta_j}{c_{i0}} \right) \right) \right) x_{ij} \tag{3.20}$$

$$= f_i(\mathbf{x}_i) \tag{3.21}$$

where $\mathbf{x}_i \triangleq \{x_{ij}, j \in M\}$ then the optimization problem can be rewritten as follows:

$$\underset{x}{minimize} : \sum_{i \in N} f_i(\mathbf{x}_i) \tag{3.22}$$

$$subject\ to : \mathbf{1}^T \mathbf{x}_i = S_i, \forall i \in N \tag{3.23}$$

$$\mathbf{1}^T \mathbf{x}_j \leq \Gamma_j, \forall j \in M \tag{3.24}$$

$$\sum_{j \in M} t_i \leq T_i, \forall i \in N \tag{3.25}$$

$$\sum_{j \in M} c_i \leq C_i, \forall i \in N \tag{3.26}$$

$$t'_i < 1, \forall i \in N \tag{3.27}$$

$$c'_i < 1, \forall i \in N \tag{3.28}$$

$$x_{ij} \geq 0, \forall i \in N, \forall j \in M \tag{3.29}$$

We then define the feasible set for the problem (3.22) as follows:

$$\mathcal{X} \triangleq \{\mathbf{x}_i | (3.23), (3.24), (3.25), (3.26), (3.27), (3.28), (3.29)\} \tag{3.30}$$

Then, following the ADMM framework, we introduce a new variable z such that

$$\begin{aligned} \underset{x}{minimize} :& \sum_{i \in N} f_i(\mathbf{x}_i) + h(z) \\ subject\ to :& \mathbf{x}_i = z \\ & \mathbf{x}_i \in \mathcal{X}, \forall i \in N \end{aligned} \tag{3.31}$$

where $h(z) = 0$ when $\mathbf{x}_i \in \mathcal{X}$.

$$h(z) = I_{\mathcal{X}}(z) = \begin{cases} 0, & \mathbf{x}_i \in \mathcal{X} \\ \infty, & otherwise \end{cases} \tag{3.32}$$

Then, the augmented Lagrangian function of (3.31) is as follows:

$$\mathcal{L}(\mathbf{x}, z, \lambda) = \sum_{i \in N} \left(f_i(\mathbf{x}_i) + \lambda_i^T (x_i - z) + \frac{\rho}{2} ||x_i - z||_2^2 \right) \tag{3.33}$$

Based on the solution from [43], the resulting ADMM variables update are the following:

$$x_i^{k+1} = \arg\min \left(f_i(x_i) + \lambda_i^{kT} (x_i - z^k) + \frac{\rho}{2} ||x_i - z^k||_2^2 \right) \tag{3.34}$$

$$z^{k+1} = \arg\min \left(h(z) + \sum_{i=1}^{N} \left(-\lambda_i^{kT} z + \frac{\rho}{2} ||x_i^{k+1} - z||_2^2 \right) \right) \tag{3.35}$$

$$\lambda_i^{k+1} = \lambda_i^k + \rho(x_i^{k+1} - z^{k+1}) \tag{3.36}$$

Algorithm 1 represents the pseudocode of our approach. Initially, we input the set of users, MEC servers, and its downlink and uplink allocation capabilities. Then, we initialize all the variables for the first iteration and fix the maximum number of iterations to 1000 runs. Moreover the Lagrangian penalty term and trade-off coefficient is set to 0.5. For each iteration, each user i updates its offloading decision. Once all users update their respective offloading decision, the MEC updates the Lagrange multiplier variable. This is followed by updating the objective function (3.10) based on the current iteration. After limited iterations, the solution converges to an optimal value.

Algorithm 1 ADMM-based task offloading

1: **input**:Initialization for $\mathcal{N}, \mathcal{M}$, **D**, **U**
2: **Output**:Minimal offloading cost
3: Initialization
4: $max\_iteration = 1000, \rho = 0.5, \alpha = 0.5$ $\mathbf{x}_i^0 \geq 0, \lambda_i^0 \geq 0, z \geq 0, t_{i0}^u, t_{i0}^d, c_{i0}, \forall i \in N$
5: **for** $k \in max\_iteration$ **do**
6: Each user $i \in \mathcal{N}$ update its offloading decision by Eqs. (3.34), (3.35), (3.32) respectively, parallelly
7: After getting updated values from all users each MEC will update λ using (3.36), parallelly
8: After all variable updated, update the objective function (3.10)
9: **end for**
return Optimal value of objective (3.10)

3.3.4 Performance Analysis

In our simulation, the MECs are assumed to be deployed at a fixed location, and N mobile users are deployed following a homogeneous Poisson point process (PPP). The input parameters of downlink and uplink bandwidth allocation follow a homogeneous uniform distribution ranging from [0 5]. In our simulations, the η value is varied from 0.4 to 0.7. Moreover, all statistical results are taken by averaging over 100 simulation runs of random location of users and bandwidth allocation. Furthermore, we also provide performance comparison by comparing our scheme (i.e., ADMM-based approach) with three other schemes. The first scheme uses a centralized method in which the optimal solution is calculated using the "convex .JL." We represent this scheme as "centralized" scheme. Then, we calculate the solution using a greedy approach in which the best first search approach is used to select the offloading decision of user to MEC. This scheme is represented as "greedy" scheme. Finally, the last scheme is the "random" scheme in which we pick random MECs for offloading its task with uniform distribution.

Figure 3.2 represents the average utility obtained by all the schemes by increasing the number of MECs in the network. It can be seen that the average utility increases with number of MEC servers. Moreover, the performance of the proposed scheme is significantly higher than the greedy and random scheme. Furthermore, the performance of the proposed ADMM scheme is indifferentiable from the centralized scheme, i.e., optimal solution under all scenarios.

Similarly, Fig. 3.3 represents the average processing cost obtained by all the schemes by increasing the number of MECs in the network. It can be inferred that the processing cost decreases as the number of MEC servers increases in the network as tasks are divided among more servers. Similarly, the performance of ADMM scheme is similar to the centralized scheme, thus obtaining an optimal solution.

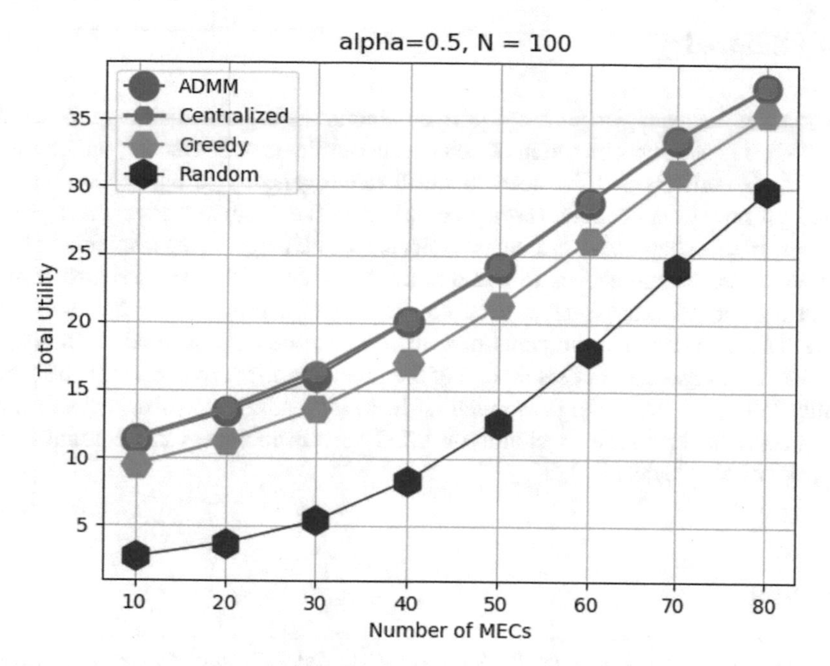

Fig. 3.2 Average utility vs. number of MEC servers

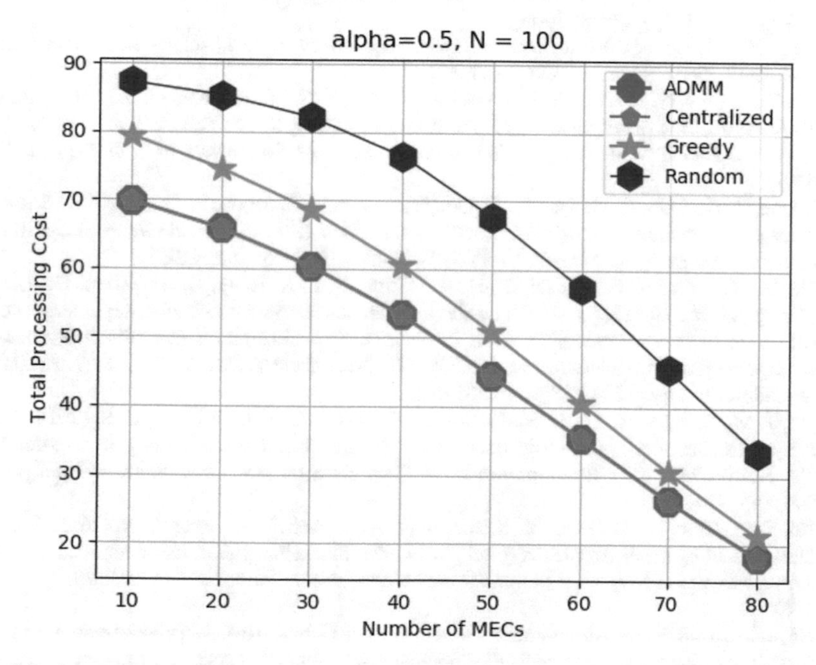

Fig. 3.3 Average processing cost vs. number of MEC servers

3.4 Summary

Radio resource management is categorized among the biggest challenge for the 5G networks due to the proliferation of heterogeneous devices. Moreover, the introduction of 5G verticals and the need to fulfill heterogeneous stringent requirements based on novel applications further complicates the radio resource management process. In this chapter, we present an overview of radio resources available in RAN and some recent approaches to manage the network. Moreover, we also discuss about network slicing which can be considered as a promising scheme to meet these heterogeneous requirements produced by various 5G verticals. Finally, we provide a use case that focuses on one of the most important requirements of virtual reality (VR), i.e., offloading computation task. Offloading intensive tasks to more resourceful devices such as clouds or MEC servers increases the computational capacity of VR devices.

References

1. Hong, C. S., Kazmi, S. A., Moon, S., & Van Mui, N. (2016). SDN based wireless heterogeneous network management. In *AETA 2015: Recent Advances in Electrical Engineering and Related Sciences* (pp. 3–12). Cham: Springer.
2. Raza, H. (2011, June). A brief survey of radio access network backhaul evolution: Part I. *IEEE Communications Magazine, 49*(6), 164–171.
3. Oo, T. Z., Tran, N. H., LeAnh, T., Kazmi, S. M. A., Ho, T. M., & Hong, C. S. (2015, August). Traffic offloading under outage QoS constraint in heterogeneous cellular networks. In *2015 17th Asia-Pacific Network Operations and Management Symposium (APNOMS)* (pp. 436–439).
4. Kazmi, S. A., Tran, N. H., Ho, T. M., Oo, T. Z., LeAnh, T., Moon, S., et al. (2015). Resource management in dense heterogeneous networks. In *2015 17th Asia-Pacific Network Operations and Management Symposium (APNOMS)* (pp. 440–443). New York: IEEE.
5. Ho, T. M., Tran, N. H., Kazmi, S. M. A., Moon, S. I., & Hong, C. S. (2016). Distributed pricing power control for downlink co-tier interference coordination in two-tier heterogeneous networks. In *Proceedings of the 10th International Conference on Ubiquitous Information Management and Communication, IMCOM '16*. New York: ACM, pp. 87:1–87:7. Available: http://doi.acm.org/10.1145/2857546.2857635
6. Ho, T. M., Tran, N. H., Le, L. B., Kazmi, S. M. A., Moon, S. I., & Hong, C. S. (2015, May). Network economics approach to data offloading and resource partitioning in two-tier LTE HetNets. In *2015 IFIP/IEEE International Symposium on Integrated Network Management (IM)* (pp. 914–917).
7. Ho, T. M., Tran, N. H., Do, C. T., Kazmi, S. M. A., LeAnh, T., & Hong, C. S. (2015, August). Data offloading in heterogeneous cellular networks: Stackelberg game based approach. In *2015 17th Asia-Pacific Network Operations and Management Symposium (APNOMS)* (pp. 168–173).
8. Ho, T. M., Tran, N. H., Ahsan Kazmi, S. M., & Hong, C. S. (2016). Distributed resource allocation for interference management and QoS guarantee in underlay cognitive femtocell networks. In *2016 18th Asia-Pacific Network Operations and Management Symposium (APNOMS)* (pp. 1–4). IEEE.

9. Cisco Visual Networking Index. (2016). *Global Mobile Data Traffic Forecast Update, 2015–2020 White Paper*. Document ID, vol. 958959758.
10. Satyanarayanan, M., Bahl, P., Caceres, R., & Davies, N. (2009, October). The case for VM-based cloudlets in mobile computing. *IEEE Pervasive Computing, 8*(4), 14–23.
11. *Fog Computing and the Internet of Things: Extend the Cloud to Where the Things Are*. (2015). Cisco White Paper.
12. Beck, M. T., Werner, M., Feld, S., & Schimper, S. (2014). Mobile edge computing: A taxonomy. In *Proceedings of the Sixth International Conference on Advances in Future Internet* (pp. 48–55). Citeseer.
13. Kazmi, S. A., Tran, N. H., Ho, T. M., & Hong, C. S. (2018). Hierarchical matching game for service selection and resource purchasing in wireless network virtualization. *IEEE Communications Letters, 22*(1), 121–124.
14. Kazmi, S. A., & Hong, C. S. (2017). A matching game approach for resource allocation in wireless network virtualization. In *Proceedings of the 11th International Conference on Ubiquitous Information Management and Communication* (p. 113). New York: ACM.
15. Kim, D. H., Kazmi, S., & Hong, C. S. (2018). Cooperative slice allocation for virtualized wireless network: A matching game approach. In *Proceedings of the 12th International Conference on Ubiquitous Information Management and Communication* (p. 94). New York: ACM.
16. Sciancalepore, V., Zanzi, L., Costa-Perez, X., & Capone, A. (2018). Onets: Online network slice broker from theory to practice. Preprint. arXiv:1801.03484.
17. Richart, M., Baliosian, J., Serrat, J., & Gorricho, J.-L. (2016). Resource slicing in virtual wireless networks: A survey. *IEEE Transactions on Network and Service Management, 13*(3), 462–476.
18. Foukas, X., Patounas, G., Elmokashfi, A., & Marina, M. K. (2017). Network slicing in 5G: Survey and challenges. *IEEE Communications Magazine, 55*(5), 94–100.
19. Zhang, H., Liu, N., Chu, X., Long, K., Aghvami, A.-H., & Leung, V. C. (2017). Network slicing based 5G and future mobile networks: Mobility, resource management, and challenges. *IEEE Communications Magazine, 55*(8), 138–145.
20. Ordonez-Lucena, J., Ameigeiras, P., Lopez, D., Ramos-Munoz, J. J., Lorca, J., & Folgueira, J. (2017). Network slicing for 5G with SDN/NFV: Concepts, architectures and challenges. Preprint. arXiv:1703.04676.
21. Ho, T. M., Tran, N. H., Kazmi, S. A., Han, Z., & Hong, C. S. (2018). Wireless network virtualization with non-orthogonal multiple access. In *NOMS 2018-2018 IEEE/IFIP Network Operations and Management Symposium* (pp. 1–9). Piscataway: IEEE.
22. Ho, T. M., Tran, N. H., Kazmi, S. M. A., & Hong, C. S. (2017, January). Dynamic pricing for resource allocation in wireless network virtualization: A Stackelberg game approach. In *2017 International Conference on Information Networking (ICOIN)* (pp. 429–434).
23. Ndikumana, A., Ullah, S., Thar, K., Tran, N. H., Park, B. J., & Hong, C. S. (2017). Novel cooperative and fully-distributed congestion control mechanism for content centric networking. *IEEE Access, 5*, 27691–27706.
24. Ndikumana, A., Ullah, S., Kamal, R., Thar, K., Kang, H. S., Moon, S. I., et al. (2015, August). Network-assisted congestion control for information centric networking. In *2015 17th Asia-Pacific Network Operations and Management Symposium (APNOMS)* (pp. 464–467).
25. Vo, P. L., Nguyen, M. N. H., Le, T. A., & Tran, N. H. (2018). Slicing the edge: Resource allocation for ran network slicing. *IEEE Wireless Communications Letters, 7*(6), 970–973.
26. Olwal, T. O., Djouani, K., & Kurien, A. M. (2016, third quarter). A survey of resource management toward 5G radio access networks. *IEEE Communications Surveys Tutorials, 18*(3), 1656–1686.
27. Rappaport, T. S., Xing, Y., MacCartney, G. R., Molisch, A. F., Mellios, E., & Zhang, J. (2017, December). Overview of millimeter wave communications for fifth-generation (5G) wireless networks—With a focus on propagation models. *IEEE Transactions on Antennas and Propagation, 65*(12), 6213–6230.

28. Ho, T. M., LeAnh, T., Kazmi, S. M. A., & Hong, C. S. (2014, September). Opportunistic resource allocation via stochastic network optimization in cognitive radio networks. In *The 16th Asia-Pacific Network Operations and Management Symposium* (pp. 1–4).
29. Adedoyin, M. A., & Falowo, O. E. (2017, June). QoS-based radio resource management for 5G ultra-dense heterogeneous networks. In *2017 European Conference on Networks and Communications (EuCNC)* (pp. 1–6).
30. Kazmi, S. M. A., Tran, N. H., Saad, W., Le, L. B., Ho, T. M., & Hong, C. S. (2016, July). Optimized resource management in heterogeneous wireless networks. *IEEE Communications Letters, 20*(7), 1397–1400.
31. Saeed, A., Katranaras, E., Zoha, A., Imran, A., Imran, M. A., & Dianati, M. (2015). Energy efficient resource allocation for 5G heterogeneous networks. In *2015 IEEE 20th International Workshop on Computer Aided Modelling and Design of Communication Links and Networks (CAMAD)* (pp. 119–123). Piscataway: IEEE.
32. Liu, G., Zhao, H., & Li, D. (2017, October). Resource allocation in heterogeneous networks: A modified many-to-one swap matching. In *2017 IEEE 17th International Conference on Communication Technology (ICCT)* (pp. 508–512).
33. Li, L., Zhao, G., & Blum, R. S. (2018, third quarter). A survey of caching techniques in cellular networks: Research issues and challenges in content placement and delivery strategies. *IEEE Communications Surveys Tutorials, 20*(3), 1710–1732.
34. Khreishah, A., & Chakareski, J. (2015, April). Collaborative caching for multicell-coordinated systems. In *2015 IEEE Conference on Computer Communications Workshops (INFOCOM WKSHPS)* (pp. 257–262).
35. Ji, J., Zhu, K., Wang, R., Chen, B., & Dai, C. (2018). Energy efficient caching in backhaul-aware cellular networks with dynamic content popularity. *Wireless Communications and Mobile Computing, 2018*, 12 pp.
36. Yang, C., Yao, Y., Chen, Z., & Xia, B. (2016, January). Analysis on cache-enabled wireless heterogeneous networks. *IEEE Transactions on Wireless Communications, 15*(1), 131–145.
37. Mao, Y., You, C., Zhang, J., Huang, K., & Letaief, K. B. (2017, fourth quarter). A survey on mobile edge computing: The communication perspective. *IEEE Communications Surveys Tutorials, 19*(4), 2322–2358.
38. Yu, S., Wang, X., & Langar, R. (2017, October). Computation offloading for mobile edge computing: A deep learning approach. In *2017 IEEE 28th Annual International Symposium on Personal, Indoor, and Mobile Radio Communications (PIMRC)* (pp. 1–6).
39. You, C., Huang, K., Chae, H., & Kim, B. (2017, March). Energy-efficient resource allocation for mobile-edge computation offloading. *IEEE Transactions on Wireless Communications, 16*(3), 1397–1411.
40. Mao, Y., Zhang, J., & Letaief, K. B. (2016, December). Dynamic computation offloading for mobile-edge computing with energy harvesting devices. *IEEE Journal on Selected Areas in Communications, 34*(12), 3590–3605.
41. Messous, M., Sedjelmaci, H., Houari, N., & Senouci, S. (2017, May). Computation offloading game for an UAV network in mobile edge computing. In *2017 IEEE International Conference on Communications (ICC)* (pp. 1–6).
42. Bastug, E., Bennis, M., Médard, M., & Debbah, M. (2017). Toward interconnected virtual reality: Opportunities, challenges, and enablers. *IEEE Communications Magazine, 55*(6), 110–117.
43. Boyd, S., Parikh, N., Chu, E., Peleato, B., & Eckstein, J. (2011, January). Distributed optimization and statistical learning via the alternating direction method of multipliers. *Foundations and Trends in Machine Learning, 3*(1), 1–122. Available: http://dx.doi.org/10.1561/2200000016

Chapter 4
Network Slicing: Radio Resource Allocation

Radio resource allocation has always been considered to be a challenging task in the cellular networks due to limited spectrum. This becomes even more challenging as fifth generation (5G) cellular networks have an even higher expectation in terms of facilitating end users with higher data rate and lower end-to-end latency. Moreover, 5G also targets to improve spectrum/energy efficiency and reduced cost for operators. One promising technology that can support to deliver is wireless network virtualization (WNV) [1, 20, 24]. Virtualization of physical resources in the data centers has played a significant role to enhance the performance by enabling abstraction and resource sharing among competing entities. Through virtualization we can reduce the operational cost of the network, provide ease of management to the operators and decoupled functionalities. WNV is a novel concept for virtualizing the radio access networks (RAN) of future cellular networks. Broadly, WNV ranges from spectrum sharing, infrastructure virtualization, to air interface virtualization. Thus, WNV abstracts the physical wireless infrastructure and radio resources to enable resource sharing among parties and enchaining the usability of the network resources [20, 24]. These resources are then isolated to a number of virtual resources (slices) and the main goal is to assign slices to different mobile virtual network operators (MVNOs) such that the network utility is maximized.

In this chapter, we discuss about the vital challenges of WNV, a number of solution approaches, and their benefits for different WNV scenarios. First, we discuss about the resource allocation problem in which one InP allocates their resources to different MVNO users. The proposed solution for the aforementioned problem allocates resources in such a manner that sum-rate of the network is maximized. Second, we discuss about resource allocation approach in a multiple InP environment. Moreover, these approaches consider spectrum resources as the OFDMA-based sub-channels. Note that in OFDMA-based scheme, the allocation will be limited by the number of spectrum resources (i.e., sub-channels, resource blocks).

S. M. A. Kazmi et al., *Network Slicing for 5G and Beyond Networks*,
https://doi.org/10.1007/978-3-030-16170-5_4

4.1 Radio Resource Allocation with Single InP

Typically, in network virtualization, the network is divided into two components: (1) the infrastructure providers (InPs) and (2) mobile virtual network operators (MVNOs). The role of an InP is to own the physical infrastructure (e.g., base station, resource blocks (RBs), backhaul, and core network) and operates the physical wireless network. On the other hand, each MVNOs will lease these InP resources to develop its own wireless network (virtual network) in order to provide services to its end users. Typically each slice is created to support a service such as VoIP, video telephony, live streaming, and video conferencing for registered users. Thus, virtualization of wireless network enables coexistence of multiples MVNOs on a shared infrastructure opposed to the traditional approach in which both the infrastructure and mobile networks were closely coupled. Through WNV this tight coupling can be removed and each InP can share its resources with multiple MVNOs (Fig. 4.1).

Efficient resource allocation will improve the resource utilization, energy efficiency, and quality of services of the network. In WNV, two kinds of resource provisioning are done when allocating resources to MVNOs: (1) rate-based provi-

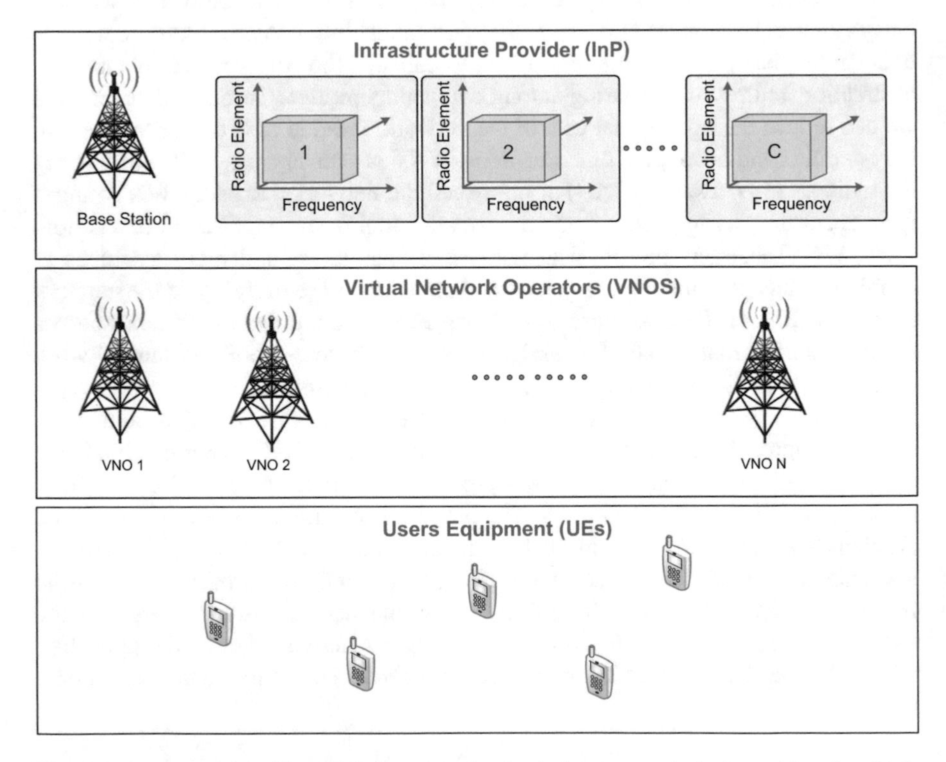

Fig. 4.1 System model with an InP that owns the physical resources (sub-channels) and multiple MVNOs

sioning in which minimum rate is guaranteed to the MVNO and (2) resource-based provisioning in which minimum number of resources are provided to the MVNOs. Moreover, there is an additional significant challenge pertaining to resource allocation in WNV, i.e., resource isolation among MVNOs. The goal is to enforce strong isolation between slices, i.e., actions in one slice do not affect another slices in the network. Thus, a number of recent resource allocation approaches in wireless network virtualization have been proposed in recent works [9, 10, 17–19, 25].

The authors in [11] designed an efficient virtual resource allocation scheme for WNV to handle the "isolation" challenge. Moreover, they propose a novel approach that considers the benefits for both the InP and MVNOs simultaneously. Two-sided matching game was employed to solve the designed allocation problem due to its special and unique ability to describe individual players roles, i.e., InP, MVNOs, and users. Matching theory is appropriate for tackling the designed problem because of the following reasons [15].

- Interactions among different type of players can be precisely described through preference profiles of players.
- Unique properties in the objective function, for example, convexity is not required to achieve an analytical tractability solution.
- Solutions for matching algorithm can guarantee key properties such as stability and optimality which makes them appropriate for online execution.

Thus, the key contribution of this work [11] can be summarized as follows:

- They formulated resource allocation in WNV as a matching game problem. To solve this problem, they also defined the preference profile for each player.
- Second, they proposed a low-complexity and stable solution for the designed resource allocation which also guaranteed the isolation constraint. Finally, they also showed that Pareto optimality is achieved by the designed solution.

4.1.1 System Model

In the system model, authors assumed a single base station (BS) owned by an infrastructure provider (InP). Moreover, the base station operated on a spectrum represented by a set $\mathcal{C}$ of orthogonal sub-channels each with bandwidth w. InP gives services based on some service level agreements (SLA) to a set $\mathcal{V}$ of mobile virtual network operators (MVNOs) and it is assumed that each MVNO v has a set of associated users as K_v. Note that an InP builds virtual resource (VR) slices based on the MVNOs demand, these slices are then allocated to MVNOs users. Generally, isolation is a primary requirement in WNV between slices of different VNOs. In this work, the authors consider that isolation between different MVNOs is achieved by guaranteeing a predetermined data rate which allows dynamic sharing of spectrum. Table 4.1 shows the notations used in this work.

Table 4.1 Table of notations with Single-InP

C_n	Set of orthogonal sub-channels
ω	Bandwidth of sub-channels
V	Set of virtual network operators
K_v	Set of users of MVNO v
$x^c_{v,k}$	Binary variable for virtual resource allocation
P^c_v	Transmit power
$\gamma^c_{v,k}$	SINR
$g^c_{v,k}$	VR gain
$r^c_{v,k}$	Achievable data rate
W	Bandwidth of VR

In their model for VR allocation, a binary variable $x^c_{v,k}$ is used:

$$x^c_{v,k} = \begin{cases} 1, & \text{if MVNOs v user } k \text{ is assigned VR } c, \\ 0, & \text{otherwise.} \end{cases}$$

Moreover, they state that one user will require only one VR. Then the following constraint holds:

$$\sum_{c \in \mathcal{C}} x^c_{v,k} \geq 1, \ \forall k \in K_v, \forall v \in \mathcal{V}. \tag{4.1}$$

Note that $x^c_{v,k} = 0$ for any user k of MVNO v with no assigned VR c. Then, the received SINR for a user k over c that has transmit power P^c_v is given by:

$$\gamma^c_{v,k} = \frac{P^c_v g^c_{v,k}}{\sum\limits_{l \in \mathcal{V}, l \neq v} P^c_l g^c_{l,k} + \sigma^2}, \tag{4.2}$$

where P^c_l is the transmit power of MVNO l user, the gain is represented by $g^c_{v,k}$ between an MVNO v and user k. Similarly, $g^r_{l,k}$ represents the gain between the MVNO l user k and MVNO v. Therefore, the achievable rate of any user k is given by:

$$r^c_{v,k} = x^c_{v,k} W \log(1 + \gamma^c_{v,k}), \tag{4.3}$$

where W is the bandwidth of VR c.

Next, the authors also study the affect for selfish players. This is considered important in WNV [8] as all players play to enhance their utility. In WNV, cost reduction is considered as the main goal, i.e., InP plays for maximizing its revenue while MVNO plays to reduce its cost that is required to be paid to the InP on receiving the VRs. Thus, InP sells its owned spectrum and charges a price, i.e.,

p_c (per unit of VR c). This spectrum is bought by the MVNOs to serve its users. The goal lies in revenue maximization for the InP by setting the appropriate price to satisfy the demand of MVNOs. Then, the network utility for an InP can be stated by:

$$U_I(\boldsymbol{x}, \boldsymbol{p}) = \sum_{v \in \mathcal{V}} x_v p_v, \tag{4.4}$$

where $x_v = \sum_{k \in K_v} \sum_{c \in \mathcal{C}} x_{v,k}^c, \forall \mathcal{V}$ represents the number of VRs required by the MVNO v, and p_v represents its price charged by InP.

On the other hand, MVNO aims to reduce its cost and serve its users in a satisfactory manner. Thus, the utility of an MVNO v is given by:

$$U_v(\boldsymbol{x}) = \sum_{k \in K_v} \sum_{c \in \mathcal{C}} r_{v,k}^c - \sum_{c \in \mathcal{C}} p_v^c \sum_{k \in K_v} x_{v,k}^c. \tag{4.5}$$

4.1.2 Problem Formulation

The objective is to augment the aggregate utility of the framework from the point of view of both InP and MVNOs. Therefore, our goal is to augment the income achieved by InP while at the same time taking care of the requests of MVNOs. Then, the VR allocation problem is given as follows:

$$\mathbf{P1}: \underset{(\boldsymbol{p},\boldsymbol{x})}{\text{maximize}}\ U_I(\boldsymbol{x}, \boldsymbol{p}) + \sum_{v \in \mathcal{V}} U_v(\boldsymbol{x}) \tag{4.6}$$

subject to:

$$\sum_{c \in \mathcal{C}} x_{v,k}^c \geq 1, \ \forall k \in K_v, \forall v \in \mathcal{V}, \tag{4.7}$$

$$r_{v,k} > r_{v,k}^{\min}, \ \forall k \in K_v, \tag{4.8}$$

$$x_{v,k}^c \in \{0, 1\}, \ \forall v, k, c, \tag{4.9}$$

$$p^{\min} \leq p_v \leq p^{\max}, \ \forall v \in \mathcal{V}. \tag{4.10}$$

In **P1**, the constraint (4.7) ensures that any user k will be given at most one VR c from the InP. Constraint (4.8) is considered as isolation provisioning constraint with minimum rate requirement for each MVNOs user. The allocation indicator is given by constraint (4.9). Finally, (4.10) the pricing constraint is presented. Problem **P1** is an integer linear programming and a combinatorial problem, and thus finding a solution incurs heavy computational complexity. Therefore it becomes challenging for a large sized network [2, 7].

The goal is to achieve a solution that has the properties of low complexity and can be implemented in a distributed fashion. To achieve such goals, matching game is utilized to convert the problem **P1** to a matching game. Next, we elaborate about the matching game used by the authors to find an efficient solution.

4.1.3 Solution

Matching theory has the capability to achieve a solution with aforementioned properties, thus it can be used for the stated allocation problem. It can overcome the challenges faced by optimization-based solution especially for combinatorial problems [6]. In matching theory, players have the ability to characterize its individual utilities relying on his local data. This helps in achieving a distributed solution that can be implemented in a self-organizing manner for the allocation problem. In this work, the authors have used the matching theory framework with two-sided preferences to address the proposed problem. This approach divides the players into two sets. Then, each member of a set ranks a subset of the individuals or members from the other set based on its local information or utility. Then, the goal of the VR allocation problem is to derive a solution for finding the best MVNO users for the all VRs that are owned by the InP. Note that by using matching theory, both sides can build their preferences to rank each other that only require local information available at each side.

Matching Game: Prerequisites

Matching theory has been successfully used in solving similar allocation problem as the stated VR allocation problem [13, 14, 16]. Based on matching theory, the authors formulate the VR allocation as a matching game problem, and design the utility of each player belonging to a specific set. Note that in this work, MVNOs users and InP owned VRs are the two distinct sets. Finally a low-complexity algorithm is presented by the authors that achieves a stable solution. Stability is a key solution concept for the matching games and will be discussed in the following subsection.

In this work from constraint (4.7), authors assume that an MVNO user k uses a single VR. However, the SLA should be met in order to use a VR, i.e., constraint (4.8). Furthermore, they consider that a VR can be allotted to multiple users until the required rate is guaranteed in order to enhance VR reuse efficiency. Thus, the designed game corresponds to a *one-to-many matching* game and is formally stated as:

$$(\mathcal{K}, \mathcal{C}, \succ_{\mathcal{K}}, \succ_{\mathcal{C}}). \tag{4.11}$$

where $\succ_{\mathcal{K}} \triangleq \{\succ_k\}_{k\in\mathcal{K}}$ states the preference ranking of a user. Similarly, $\succ_{\mathcal{C}} \triangleq \{\succ_c\}_{c\in\mathcal{C}}$ states the VRs. Then, a matching μ is defined formally as follows.

Definition 4.1 A *matching* μ is defined by a function from the set $\mathcal{K} \cup \mathcal{N}$ into the set of elements of $\mathcal{K} \cup \mathcal{N}$ such that:

1. $|\mu(k)| \leq 1$ and $\mu(k) \in \mathcal{N}$,
2. $|\mu(n)| \leq q_n$ and $\mu(n) \in \mathcal{K} \cup \phi$,
3. $\mu(k) = n$ if and only if k is in $\mu(n)$,

where $\mu(k) = \{n\} \Leftrightarrow \mu(n) = \{k\}$ for $\forall n \in N, \forall k \in K$ and $|\mu(.)|$ represents the cardinality of the outcome produced by the matching $\mu(.)$. The first two conditions express that the matching type, i.e., one-to-many type, i.e., a UE k can join only one InP n whereas an InP n can accommodate multiple UEs up to q_n, given the SLA is guaranteed from the constraints (4.7) and (4.8) of **P1**. q_c denotes the set of users that can be admitted on a VR c (i.e., quota) such that the required SLA is met by using the VR. Furthermore, there are additionally situations in which user k is not suitable for any VR c due to breaching of requirement constraint (4.7) which is represented by $\mu(c) = \phi$.

Matching Game: Preference Profile

To build the preference profile, interaction between the InP and MVNOs is required. It is assumed by the authors that the InP broadcasts its desired price for a VR initially. Then, each MVNOs also responds by stating its demand based on the users associated with an MVNO. As stated earlier, each MVNOs goal is to provide service to its users at the minimum price. On the other hand, InP aims to maximize its revenue by setting the right price to satisfy the demand of MVNOs while guarantees the contracts agreements, i.e., minimum required bandwidth.

A preference profile is built by each player of both sides to assess one another. In the proposed game, local information will be used by each side to build their respective preference profiles. For the MVNO users, the preference profile is built using the following preference function:

$$U^k(c) = \max\{(r^c_{v,k} - r^{\min}_{v,k}), 0\} \tag{4.12}$$

This function represents the aim that an MVNO wants to achieve, i.e., provide best service at least cost that is required to be paid to InP. Equation (4.12) also reflects the feasibility condition as stated in constraint (4.8). It states that in case a VR c is unable to fulfill the demand requirement for a user k, then a zero value is given by (4.12) and is not considered in the preference profile or ranking. Note that each MVNO V will select its best user k using the following function on a VR c.

$$U^v(c) = \max_k U^k(c) \tag{4.13}$$

Moreover, a user k builds its preference profile $\mathcal{P}_k$ by ranking the set of VRs in an increasing order, i.e., the cheaper the cost of VR, the better its position in ranking which implies, a VR $c \in \mathcal{C}$ which has the lowest cost through (4.12) will be the first preferred VR c from the set of VRs, i.e., $c' \in \mathcal{C}$. For a user k this is given by:

$$c \succ_k c', \ \ \forall c,\ c' \in \mathcal{C}. \tag{4.14}$$

Thus user k will like to use VR c for performing its transmission. Finally, it is to be noted that preference profiles P_k for all users k are maintained by their respective MVNOs.

Similarly, a VR c also requires a preference profile to rank users $k \in K_v$. InP helps to build this preference profile on behalf of each VR c which is given by $\mathcal{P}_c$. It is given by the following function:

$$U^c(\mathcal{N}) = \max_{k_v}\{|\mathcal{N}_{k_v}| : r^c_{\mathcal{N}_{k_v}} > r^{\min}_{\mathcal{N}_{k_v}}\}, \tag{4.15}$$

where $\mathcal{N}$ denotes the set of accepted users k associated with any MVNO v which can be allocated by the VR c while maintaining the SLA as follows:

$$\mathcal{N} \subseteq \bigcup\nolimits_{v \in \mathcal{V}} K_v, \tag{4.16}$$

Similarly, $r^c_{\mathcal{N}_{k_v}}$ denotes the rate for these users, and $r^{\min}_{\mathcal{N}_{k_v}}$ represents the minimum requirement of all users.

As indicated by (4.15), a VR c picks multiple users $\mathcal{N}$ to form a subset with the end goal that achievable data rate accomplished by the subset of $\mathcal{N}$ is more than the SLA, i.e., required rate of all users. The goal of the designed function is to maximize to choose a set with maximum number of users in $\mathcal{N}$. Note this enables the InP to produce the greatest income (4.4) while at the same time abiding the SLAs between MVNOs and InP. The largest subset in terms of number of members is ranked highest and is the most favored among all the possible subsets.

In order to calculate the preference profiles of all users, communication of some information is required in the network. For the InP side, it requires to know the minimum required data rate for each MVNO user, which can be sent with some signaling through the control channel. Similarly, the MVNOs need to know the prices of the VRs to evaluate all available VRs. Nonetheless, signaling just associated with sending these values once from the VNOs to the InP, which is relatively small [23]. After building the preference profiles by each side, next, the aim is to design an efficient solution to match players of both sides with each other.

Matching Game: VR Allocation Scheme

In this section, the authors present the proposed VR allocation scheme based on the aforementioned game. The goal of the proposed algorithm is to find a stable allocation, i.e., key solution concept in matching theory [5, 22]. Then, the stability of matching game is defined formally as follows.

Algorithm 1 Resource allocation algorithm

1: ***Phase 1: Initialization***:
2: **input**: $\mathcal{P}_k, \mathcal{P}_c, \forall c, k$.
3: **initialize:** $t = 0$, $\mu^{(t)} \triangleq \{\mu(k)^{(t)}, \mu(c)^{(t)}\}_{k \in \mathcal{K}, c \in \mathcal{C}} = \emptyset$, $\mathcal{L}_c{}^{(t)} = \emptyset$ $\mathcal{P}_k{}^{(0)} = \mathcal{P}_k$, $\mathcal{P}_c{}^{(0)} = \mathcal{P}_c$, $\forall c, k$.
4: ***Phase 2: Matching***:
5: **repeat**
6: $t \leftarrow t + 1$
7: **for** $c \in \mathcal{C}$, propose k according to $\mathcal{P}_c{}^{(t)}$ **do**
8: **while** $c \notin \mu(k)^{(t)}$ and $\mathcal{P}_c^{(t)} \neq \emptyset$ **do**
9: **if** $U^k(c) \geq 0$, **then**
10: **if** $c \succ_k \mu(k)^{(t)}$ **then**
11: $\mu(k)^{(t)} \leftarrow \mu(k)^{(t)} \setminus \mu(k)^{(t-1)}$
12: $\mu(k)^{(t)} \leftarrow c$
13: $\mathcal{P}'^{(t)}_c = \{j' \in \mu(k)^{(t-1)} | c \succ_k j'\}$
14: **else**
15: $\mathcal{P}''^{(t)}_c = \{c \in \mathcal{C} | \mu(k)^{(t)} \succ_k c\}$
16: **end if**
17: **else**
18: $\mathcal{P}'''^{(t)}_c = \{c \in \mathcal{C} | U^k(c) \leq 0\}$
19: $\mathcal{L}_c{}^{(t)} = \{\mathcal{P}'^{(t)}_c\} \cup \{\mathcal{P}''^{(t)}_c\} \cup \{\mathcal{P}'''^{(t)}_c\}$
20: **for** $l \in \mathcal{L}_c{}^{(t)}$ **do**
21: $\mathcal{P}_l{}^{(t)} \leftarrow \mathcal{P}_l{}^{(t)} \setminus \{k\}$
22: $\mathcal{P}_k{}^{(t)} \leftarrow \mathcal{P}_k{}^{(t)} \setminus \{l\}$
23: **end for**
24: **end if**
25: **end while**
26: **end for**
27: **until** $\mu^{(t)} = \mu^{(t-1)}$
28: ***Phase 3: Resource Allocation***:
29: **output**: $\mu^{(t)}$

Definition 4.2 A matching μ is stable if there exists no blocking pair (k, c), where $k \in \mathcal{K}, c \in \mathcal{C}$, such that $c \succ_k \mu(k)$ and $k \succ_c \mu(c)$, where $\mu(k)$ and $\mu(c)$ represent, respectively, the current matched partners of k and c.

The output of the algorithm is the assignment vector of users and the pseudocode is given in Algorithm 1. The proposed algorithm is ensured to converge to a stable allocation as it is a variation of "deferred-acceptance algorithm" [22].

The algorithm has three stages, namely *the introduction stage, the matching stage*, and the *VR allocation stage*. In the *initialization stage*, local information is acquired by both the sides to build preference profile and perform the ranking operation. It begins by each MVNO presenting its demand for each user, i.e., $r^{\min}_{v,k}$, $\forall K_m, \mathcal{V}$. In light of it, the InP communicates its available resources and the relating costs for utilization. In response to this, each MVNO user builds its preference profile $\mathcal{P}_k$. So also, the InP builds its profile for each VR c, i.e., $\mathcal{P}_r$ for all the MVNO users (lines 1–3). In stage two *matching stage*, the best MVNO user k receives a proposal for each VR c as indicated by its preference profile $\mathcal{P}_k$

(lines 7–8). This results in either of the two cases, case 1: If (4.12) is fulfilled, i.e., MVNO user k has a nonnegative utility for VR c. At that point, user k accepts the proposal temporarily, if its unmatched, i.e., $\mu(k)^{(t)} = \phi$ (lines 9–12). In the event that its already matched $\mu(k)^{(t)} \neq \phi$, then a comparison is made between the current match and the new proposal. The proposal that is higher positioned in its preference profile is accepted while the lower positioned is rejected (line 13 and 15). In case 2, the proposing VR c will be dismissed as it doesn't meet (4.12). All the rejected VRs will at last be included in the rejected list, i.e., the set $\mathcal{L}_c{}^{(t)}$, at iteration t (Line 19). Then, using the rejected list all players update their preference profile. This implies the removal of all MVNO users k from the $\mathcal{P}_c{}^{(t)}$ by VR c, and similarly MVNO users also remove c from $\mathcal{P}_k{}^{(t)}$ (lines 20–22). Note that the matching process is an iterative process which is completed when the solution is found, i.e., a stable solution between the two sides (line 27). The process converges when two consecutive matching remain same (line 27). The last stage is the *VR allocation* stage, in this stage the matched MVNO users are permitted to use on the apportioned VRs (lines 28–29). Finally, the authors also provide the convergence and proof for Pareto optimality as follows.

Proposition 4.1 *Algorithm 1 converges to a stable point as it is variant of deferred acceptance algorithm [22].*

Proof The proof of Theorem 4.1 can be found in [[11], Theorem 1].

Similarly, they also define the Pareto optimality as follows.

Definition 4.3 A matching μ results in a weak Pareto optimal solution if there exists no other matching μ' that can achieve a better utility, where the inequality is component-wise and strict for a pair (k, c).

Theorem 4.1 *Algorithm 1 produces a weak Pareto optimal solution for the InP [11].*

Proof The proof of Theorem 4.1 can be found in [[11], Theorem 2].

4.1.4 Simulation Setup and Results

In this section, simulations results performed by authors are presented. They carry out simulations under different topologies and situations to exhibit the execution and adequacy of the VR allocation algorithm. The network topology includes one BS owned by an InP, with MVNO users randomly located inside a circle of radius of $r = 500$ m. For their simulations, they considered a network with only 3 MVNOs each with 5 associated users as shown in Fig. 4.2. Furthermore, they cater two cases when the InP possesses $C = 5$ and 10 OFDMA sub-channels, every one of which has an aggregate data transfer capacity of 180 KHz. Moreover, detailed wireless system parameters such as channel gains and BS power are stated in Table 4.2 [11].

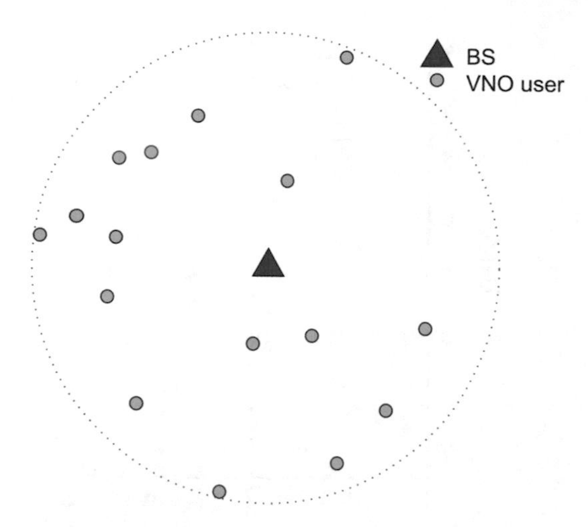

Fig. 4.2 Simulation topology

Table 4.2 Default simulation parameters with Single-InP

Simulation parameters	Values
Coverage area	$r = 500$ m
Number of VNO	3
Number of subscribed users in each VNO	5
Bandwidth per sub-channel	180 KHz
Rayleigh random variables	Unit variance
Path loss component	$\beta = 3$
MVNOs minimum data rate	$r_{\min} = 5$ bps/Hz/user
Maximum BS transmission power	43 dBm

Additionally, the presented results are averaged over a large number of independent simulation runs.

In Fig. 4.3, the required number of iteration to achieve convergence of Algorithm 1 is presented for different number of users in the system. Algorithm 1 achieves the convergence point when both the sides reach a stable point and don't plan to stray from their present allocation. It can be deduced from the graph that under all situations, the scheme achieves convergence after limited number of iterations, i.e., less than 4 for the two situations when $C = 10$ and 5 with 10 users. Besides, take note of that, for a lesser number of VRs in the network, the convergence is achieved earlier. The reason behind this is that VRs are limited and no more users are allowed to be admitted to abide constraint (4.8).

In Fig. 4.4, the achievable throughput is presented versus the number of users in the network. To perform this simulation, number of users in the network were increased to observe the average throughput. It was found that the throughput increases with users linearly until enough number of users are not there in the system. When the number of users in the system become adequately vast especially for the case of $C = 5$, constraint (4.8) needs to be protected by allowing only some

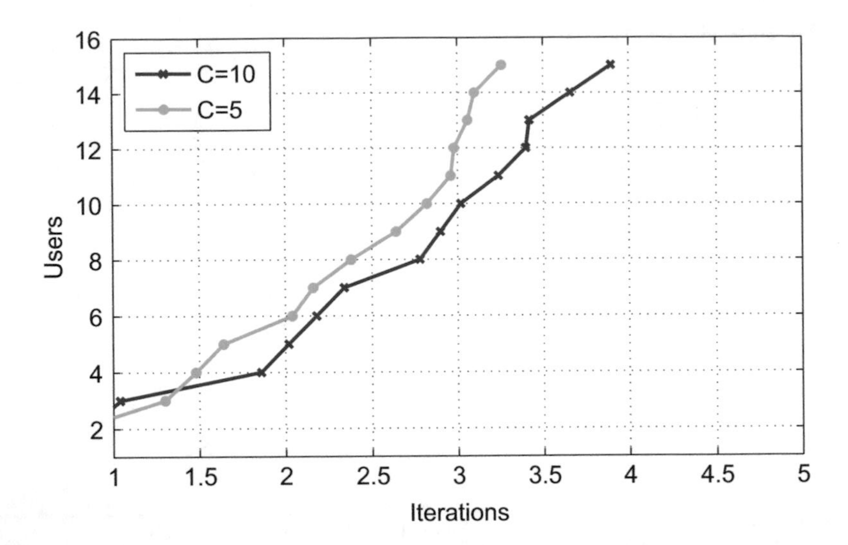

Fig. 4.3 Number of users vs. required iterations

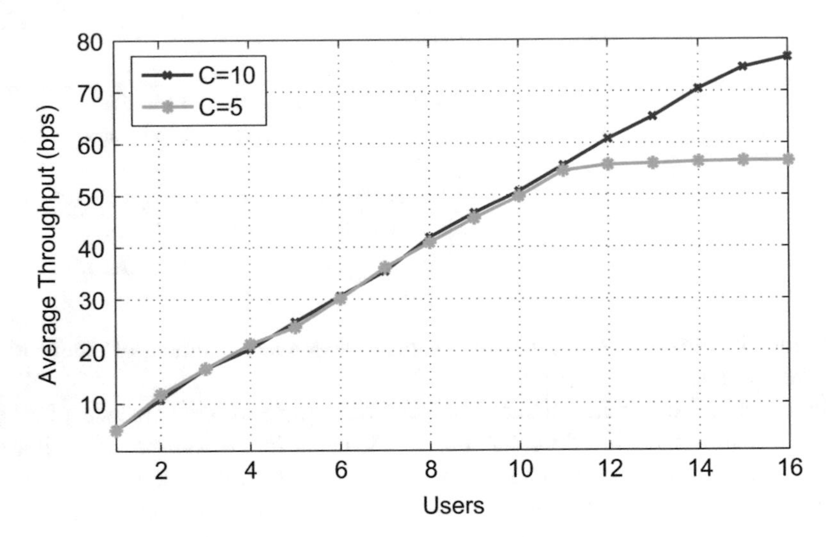

Fig. 4.4 Average throughput vs. number of users

users which results in saturation of throughput. Besides, for the situation when there exists more VRs, i.e., 10, the throughput increments further. Note that the revenue of an InP increases with more number of users.

This work proposed an efficient resource allocation scheme for the MVNO users by employing the matching theory. They formulate the resource allocation problem and propose a stable and decentralized solution for the proposed problem.

Stability was proved and it was shown that the scheme generates a Pareto optimal solution. However, the proposed model assumed static demands and prices and did not consider dynamic isolation and prices in this work. Next, we present another scheme in [12] in which dynamic demands and prices are considered.

4.2 Radio Resource Allocation with Multi-InP

In the previous section, we discussed about the resource allocation in a single InP environment. However, in a practical system, we have a group of InPs available in a service area that can serve a group of MVNOs and its users. In a multi-InP environment, these InPs can enhance the resource sharing by servicing MVNOs by giving them isolated resources (i.e., slices) formed via the physical owned resources. These virtual resources can be shared among multiple MVNOs which can enhance the flexibility of a service area. The challenge in such a setting is to perform resource allocation in such a manner that the performance of the service area consisting of multiple InPs and MVNOs is enhanced. Some related works on this specific issue can be found in [3, 21, 26].

The work in [12] also addresses the problem of resource allocation in a specific service area that contains multiple InPs. In this work, service selection and resource purchasing is performed in a multi-InP environment. Service selection allows a number of users to be associated with an MVNO and resource purchasing requires the resources of an InP to be purchased by the MVNOs to serve its associated users.

In summary, the key novelty and contributions include:

- They formulate the *service selection* and *resource purchasing* problems for a multi-cell WNV. Moreover, the problem considers utility maximization for both the InPs and MVNOs subject to the service contract agreements between InPs and MVNOs.
- They develop a low-complexity algorithm called the hierarchal matching algorithm which can achieve a near-optimal solution.

4.2.1 System Model

The authors consider a service area that contains a set of N base stations (BSs) which are owned by different MVNOs. Moreover, in their model, they assume that each physical resources is orthogonal to other InPs, i.e., no inter InP interference is considered. Furthermore, there are M MVNOs that require service from the InPs in the service area. These MVNOs then provide service to a set of K_m users (UEs). Figure 4.5 illustrates the system model (Table 4.3).

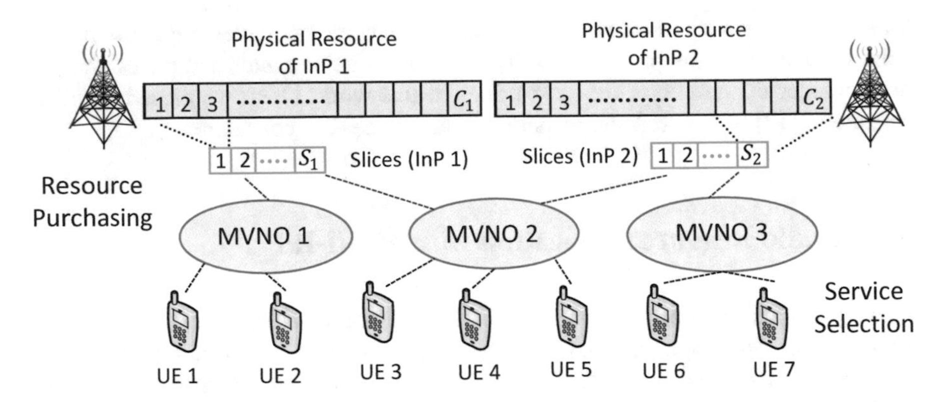

Fig. 4.5 System model: the InP owns the physical resources, virtualizes them into slices, and allocates to multiple MVNOs

Table 4.3 Table of notations with Multi-InP

N	Number of InP-base stations
M	Number of MVNOs
K	Total number of user equipment
$x_{k,m}$	Binary service selection variable between UE and MVNO
$y_{m,n}^{Sn}$	Binary resource purchasing decision of InP
s_n	Number of slices
m	Proposal purchased from MVNO
d_m	UE demand of the MVNO
ω	Weight characterizing trade-off b/w fairness and InPs revenue
K_m	Subscribed user equipment
C_n	Set of orthogonal channels
W	Bandwidth of each channel
σ^2	Background noise
P_n	Power on each channel
$P_n^{\max}$	Maximum power of an InP
S_n	Isolated services by InPs
$R_{n,k}^{Sn}$	Data rate for a UE
$\gamma_{n,k}^{c}$	Channel gain between InP-BS n and UE k on channel c
β_m^M	Per unit price of MVNO
d_k	Demand of UE

In this model, every InP has C_n orthogonal set of channels having bandwidth W. They consider a system with no inter-InP interference and thus only noise is considered on a channel which is represented by σ^2. Moreover, they consider that an InP n divides its power equally on all channels, i.e., $P_n = \frac{P_n^{\max}}{|C_n|}$, where P_n and $P_n^{\max}$ represent the power on each channel and the maximum power of an InP n, respectively. Moreover, each InP n provides services to MVNOs through S_n set of slices, where each slice s_n is different and is based on MVNOs m demand. Then, the throughput achieved by a UE k on a specific slice s_n of an InP n is given as follows:

$$R_{n,k}^{s_n} = \sum\nolimits_{c \in s_n} W \log\left(1 + \gamma_{n,k}^{c}\right), \tag{4.17}$$

where $\gamma_{n,k}^{c} = \frac{P_n g_{n,k}^{c}}{\sigma^2}$.

4.2.2 Problem Formulation

The goal in this work is to maximize the performance in a service area consisting of UEs, MVNOs, and InPs. Therefore, for each UE $k \in K$ its problem is defined as follows:

$$\textbf{UE}: \min_{x_{k,m} \in \{0,1\}} \sum\nolimits_{m \in M} x_{k,m} \beta_m^M d_k, \tag{4.18}$$

$$\text{s.t.} \sum\nolimits_{m \in M} x_{k,m} = 1, \tag{4.19}$$

where $x_{k,m} \in \boldsymbol{X}$ and $x_{k,m} = 1$ represents the UE k wants to buy service from MVNO m for its demand d_k. Therefore, each UE aims on minimizing (4.18) for its demand. Moreover, constraint (4.19) ensures that a UE is served by one MVNO.

Then, an MVNO m aims to maximize its benefits by buying resources for UEs and determining the required channels at minimum cost. The MVNO m problem is given as:

$$\textbf{MVNO}: \max_{\tilde{x}_{k,m}, \tilde{y}_{m,n}^{s_n} \in \{0,1\}} \sum\nolimits_{k \in K} \tilde{x}_{k,m} \beta_m^M d_k -$$

$$\sum\nolimits_{n \in N} \sum\nolimits_{s_n \in S_n} \tilde{y}_{m,n}^{s_n} \beta_n^I |s_n|, \tag{4.20}$$

$$\text{s.t.} \sum\nolimits_{m \in M} \tilde{x}_{k,m} \leq 1, \ \forall k, \tag{4.21}$$

$$\sum\nolimits_{k \in K} \tilde{x}_{k,m} l_{k,n} \leq \tilde{y}_{m,n}^{s_n} |s_n|, \ \forall n, \tag{4.22}$$

where $\tilde{x}_{k,m} \in \tilde{\boldsymbol{X}}$ and $\tilde{y}_{m,n}^{s} \in \tilde{\boldsymbol{Y}}$ represent the service selection and resource purchasing variables, respectively. Formally, they are given as follows:

$$\tilde{x}_{k,m} = \begin{cases} 1, & \text{if UE } k \text{ is selected by MVNO } m \text{ for service,} \\ 0, & \text{otherwise.} \end{cases}$$

$$\tilde{y}^{s}_{m,n} = \begin{cases} 1, & \text{if MVNO } m \text{ aims to buy slice } s_n \text{ of InP } n, \\ 0, & \text{otherwise.} \end{cases}$$

Here, $l_{k,n}$ represents the required channels for ensuring the demand d_k. Finally, constraint (4.21) represents that k is associated with one MVNO and constraint (4.22) restricts an InP n to over-allocate resources.

Then, the InP solves the following problem to maximize its benefit:

$$\mathbf{InP}: \max_{y^{s_n}_{m,n} \in \{0,1\}} \sum_{m \in M} \sum_{s_n \in S_n} y^{s_n}_{m,n} \Big(\sum_{k \in K_m} \log(R^{s_n}_{n,k}) + \omega \beta^{I}_{n} |s_n| \Big) \tag{4.23}$$

$$\text{s.t.} \sum\nolimits_{m \in M} \sum\nolimits_{s_n \in S_n} y^{s_n}_{m,n} \leq |S_n|, \tag{4.24}$$

$$\sum\nolimits_{k \in K_m} \sum\nolimits_{s_n \in S_n} y^{s_n}_{m,n} R^{s_n}_{n,k} \geq d_m, \ \forall m, \tag{4.25}$$

where $y^{s_n}_{m,n} \in \boldsymbol{Y}$ is the InP n decision variable and is given as follows:

$$y^{s_n}_{m,n} = \begin{cases} 1, & \text{if MVNO } m \text{ is selected by InP } N \text{ for resource purchasing,} \\ 0, & \text{otherwise.} \end{cases}$$

Moreover, the InP sells resource based on the requirement of an MVNO which is represented by the demand d_m from the MVNO m (i.e., $d_m = \sum_{k \in K_m} d_k$). Furthermore, (4.23) represents the objective of InP in terms of its benefits. Constraint (4.24) protects the InP to perform the resource violation and the constraint in (4.25) represents the contract agreement that is considered between MVNO and InP. This is also considered as the isolation constraint in the WNV.

The designed optimization problem turns out to be a mix integer linear programming problem [2] and solving this problem becomes challenging due to its combinatorial nature. To solve this problem, the authors have designed a hierarchical matching mechanism which can solve the hierarchical (i.e., two-level) WNV problem.

4.2.3 Solution

In the hierarchical matching mechanism, two levels are considered. In the first level (low-level), UE and MVNO play a matching game to find the UEs that are successful in service selection, while in the second level (high-level) MVNO and InP decide the resource purchasing. Two games using two-sided matching games are designed to achieve this objective. In the high-level, InP and MVNOs act as the two sides of the matching game and represent the vendor and buyers in this game, respectively. In the lower level, MVNOs and UEs play the second game in which the MVNOs play the vendors' role and UEs act as the buyers in this game. As a one-to-many matching game is formulated, then, a buyer can be associated with a single vendor while vendor can have multiple buyers matched to it. Formally, it can be given by the tuple $(B, V, q_v, \succ_B, \succ_V)$. Where $\succ_B \triangleq \{\succ_b\}_{b \in B}$ and $\succ_V \triangleq \{\succ_v\}_{v \in V}$ denote the preference relation sets of the buyers B and vendors V, respectively.

Definition 4.4 A *matching* μ is defined by a function from the set $B \cup V$ into the set of elements of $B \cup V$ such that:

(1) $|\mu(b)| \leq 1$ and $\mu(b) \in V$,
(2) $|\mu(v)| \leq q_v$ and $\mu(v) \in 2^{|B|} \cup \phi$, where q_v is the quota of v,
(3) $\mu(b) = v$ if and only if b is in $\mu(v)$.

In the hierarchical matching mechanism, first, the low-level game is solved to find the set of UEs that will be served by the MVNOs. Then, the high-level game is played between the InPs and the MVNOs. In the low-level game, the UEs rank all MVNOs based on the their local information which is reflected by the following preference function to build its preference profile $\mathcal{P}_k^l$.

$$U_k^{b_l}(m_n) = \beta_{m_n}^M d_k, \ \forall m_n. \tag{4.26}$$

Similarly, each MVNO m_n also build its preference profile $\mathcal{P}_{m_n}^l$ by ranking all UEs based on the following function:

$$U_{m_n}(k) = \max(\beta_{m_n}^M d_k - \beta_n^I l_{k,n}, 0), \ \forall k. \tag{4.27}$$

Note that, here, MVNO m_n requires to calculate the required channels by UE k $l_{k,n}$ to fulfill its demand d_k. Then, the goal is to find a service selection scheme via matching.

After solving the low-level game, the high-level game is played for resource purchasing. To evaluate the two sides in this game, here also preferences profiles are built by the two sides, i.e., MVNOs and INPs given by $\mathcal{P}_{m_n}^u$ and $\mathcal{P}_n$, respectively. The MVNOs use the following function to rank InPs in this game.

$$U_{m_n}^{b_h}(n) = \beta_n^I d_{m_n}, \ \forall n. \tag{4.28}$$

where d_{m_n} is the demand on each MVNO m posed by its accepted UEs for an InP n and given by

$$d_{m_n} = \sum_{k \in \mu(m_n)} d_k. \tag{4.29}$$

Finally, the InP also ranks all MVNOs by using the following function. The goal is to maximize its benefit by allocating its slices to MVNOs

$$U_n^{Sh}(m_n) = \sum\nolimits_{k \in \mu(m_n)} \log(R_{n,k}^{S_n}) + \omega \beta_n^I \gamma_{m_n}, \ \forall m_n. \tag{4.30}$$

Hierarchical Matching (HM) Game Algorithm

Once the preferences profile is built, the nest goal is to develop a stable matching algorithm for the designed hierarchical problem. However, designing a solution for a hierarchical game is challenging as more strict conditions need to be followed in order to achieve stability. Therefore, in this work, the authors have proved not only the stability in each stage of the hierarchal structure but also a concept of group stability which is a more strong condition when achieving stability.

The main challenge in a hierarchical game is that any change in the low-level player's strategy will affect the strategy of the players in the high-level. Therefore, convergence and stability cannot be achieved until the player's strategy at the low-level is fixed. This is achieved by forming a group and achieving a group stability for every InP n, i.e., $\mathcal{G}_n, \forall n \in N$. Moreover, the game is played in a sequential manner as shown in Fig. 4.6. Finally, the detailed pseudocode of the novel stable matching algorithm is presented in Algorithm 2 by the authors.

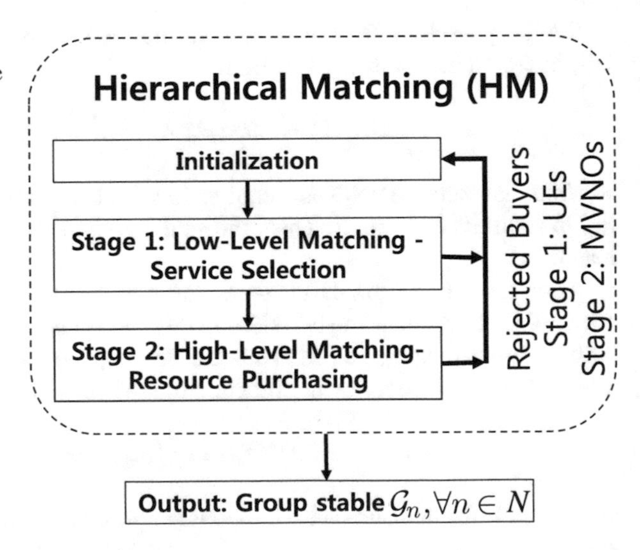

Fig. 4.6 Block diagram of hierarchical matching scheme

Algorithm 2 Hierarchal RA algorithm (HM)

1: **initialize:** $\tau = 0, \mathcal{G}_n^\tau = 0, \forall n$.
2: **while** $\mathcal{G}_n^\tau \neq \mathcal{G}_n^{\tau+1}$ **do**
3: $\tau = \tau + 1$.
Stage 1: Low-Level Matching - Service Selection:
4: **input**: $t = 0, q_{m_n}^{(0)} = q_{m_n}^\tau, \mathcal{P}_k{}^{(0)} = \mathcal{P}_k, \mathcal{P}_{m_n}{}^{(0)} = \mathcal{P}_{m_n}^l, \forall m_n, k \notin \mathcal{G}_n^\tau$.
5: $t \leftarrow t + 1, \forall k \in \mathcal{K}$, propose to m_n according to $\mathcal{P}_k{}^{(t)}$.
6: **while** $k \notin \mu(m_n)^{(t)}$ and $\mathcal{P}_k^{(t)} \neq \emptyset$ **do**
7: **if** $q_{m_n}^{(t)} \leq l_{k,n}$ **then**
8: $\mathcal{P'}_{m_n}^{(t)} = \{k' \in \mu(m_n)^{(t)} | k \succ_{m_n} k'\} \cup \{k\}$.
9: $k'_{lp} \leftarrow$ the least preferred $k' \in \mathcal{P'}_{m_n}^{(t)}$.
10: **while** $(\mathcal{P'}_{m_n}^{(t)} \neq \emptyset) \cup (q_{m_n}^{(t)} \geq l_{k,n})$ **do**
11: $\mu(m_n)^{(t)} \leftarrow \mu(m_n)^{(t)} \setminus k'_{lp}, \quad \mathcal{P'}_{m_n}^{(t)} \leftarrow \mathcal{P'}_{m_n}^{(t)} \setminus k'_{lp}$.
12: $q_{m_n}^{(t)} \leftarrow q_{m_n}^{(t)} + l_{k'_{lp},n}, \quad k'_{lp} \leftarrow k' \in \mathcal{P}'_{m_n}{}^{(t)}$.
13: **end while**
14: Remove rejected players from $\mathcal{P}_k{}^{(t)}$ and $\mathcal{P}_{m_n}{}^{(t)}$.
15: **else**
16: $\mu(m_n)^{(t)} \leftarrow \mu(m_n)^{(t)} \cup \{k\}, \quad q_{m_n}{}^{(t)} \leftarrow q_{m_n}{}^{(t)} - l_{k,n}$.
17: **end if**
18: **end while**
19: $\tilde{X} \leftarrow \mu^*$
Stage 2: High-Level Matching- Resource Purchasing:
20: **input**: $t = 0, q_n^{(0)} = q_n^\tau, \mathcal{P}_{m_n}{}^{(0)} = \mathcal{P}_{m_n}^u, \mathcal{P}_n{}^{(0)} = \mathcal{P}_n, \forall m_n, n$.
21: $t \leftarrow t + 1, \forall m_n$, propose to n according to $\mathcal{P}_{m_n}{}^{(t)}$.
22: **while** $m_n \notin \mu(n)^{(t)}$ and $\mathcal{P}_{m_n}^{(t)} \neq \emptyset$ **do**
23: **if** $q_n^{(t)} \leq |\gamma_{m_n}|$ **then**
24: $\mathcal{P'}_n^{(t)} = \{m_n' \in \mu(n)^{(t)} | m_n \succ_n m_n'\} \cup \{m_n\}$.
25: $m_n'{}_{lp} \leftarrow$ the least preferred $m_n' \in \mathcal{P'}_n^{(t)}$.
26: **while** $(\mathcal{P'}_n^{(t)} \neq \emptyset) \cup (q_n^{(t)} \geq |\gamma_{m_n}|)$ **do**
27: $\mu(n)^{(t)} \leftarrow \mu(n)^{(t)} \setminus m_n', \quad \mathcal{P'}_n^{(t)} \leftarrow \mathcal{P'}_n^{(t)} \setminus m_n'{}_{lp}$.
28: $q_n^{(t)} \leftarrow q_n^{(t)} + |\gamma_{m_n'{}_{lp}}|, \quad m'_{n_{lp}} \leftarrow m_n' \in \mathcal{P}'_n{}^{(t)}$.
29: **end while**
30: Remove rejected players from $\mathcal{P}_{m_n}{}^{(t)}$ and $\mathcal{P}_n{}^{(t)}$.
31: **else**
32: $\mu(n)^{(t)} \leftarrow \mu(n)^{(t)} \cup \{m_n\}, \quad q_n{}^{(t)} \leftarrow q_n{}^{(t)} - |\gamma_{m_n}|$.
33: **end if**
34: **end while**
35: $Y \leftarrow \mu^*$
36: Update $\mathcal{G}_n^\tau, \forall n$.
37: **end while**
38: **output**: Convergence to group stable $\mathcal{G}_n, \forall n$.

4.2.4 *Performance Analysis*

To simulate the proposal, the authors consider the standard parameters of cellular technologies following the system guidelines [4]. Moreover, other parameters used for this simulation are described in Table 4.4. Furthermore, the presented results are averaged over 500 number of independent simulation runs.

For comparison purposes, the proposed approach was evaluated against two schemes [26]. The first scheme was a fixed sharing scheme (FS), in which equal number of channels are reserved for all MVNOs. The second scheme is a general sharing scheme (GS). In this scheme, each InP performs channel allocation for all users in the network. Such a scheme has been used for channel allocation in some existing works such as [3, 21].

In Fig. 4.7, the authors calculated the average sum-rate for different network size under different schemes. It can be inferred that sum-rate increases with the number of users in the system. This however saturates as the number of users in the system increases as each InP has limited channel band (i.e., 1.4 MHz). Moreover, the proposed scheme closely performs when compared to the GS scheme and achieves

Table 4.4 Default simulation parameters with Multi-InP

Simulation parameters	Values
Coverage area	1000 * 1000 m
Number of MVNO in network	5
Bandwidth of each InP	1.4 MHz
Bandwidth of each channel	Normalized value 1
Weight parameter ω	Normalized value 1
UE demand	1–3 bps/Hz
Prices for MVNOs	β_m^M = 4–8 monetary units/bps/Hz
Prices for InPs	β_n^I = 2–4 monetary units/bps/Hz

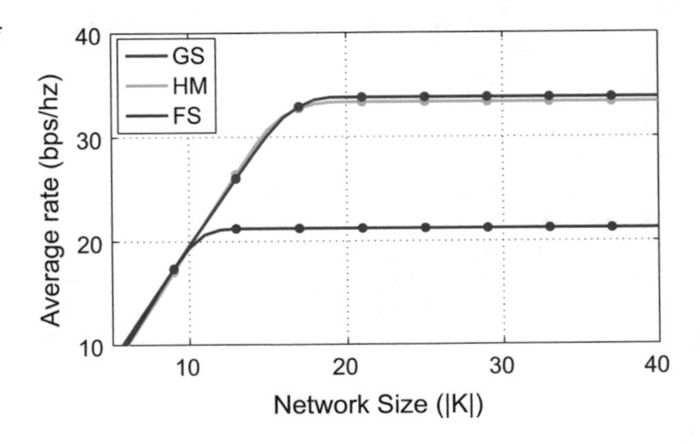

Fig. 4.7 Average sum-rate of HM, GS, and FS schemes

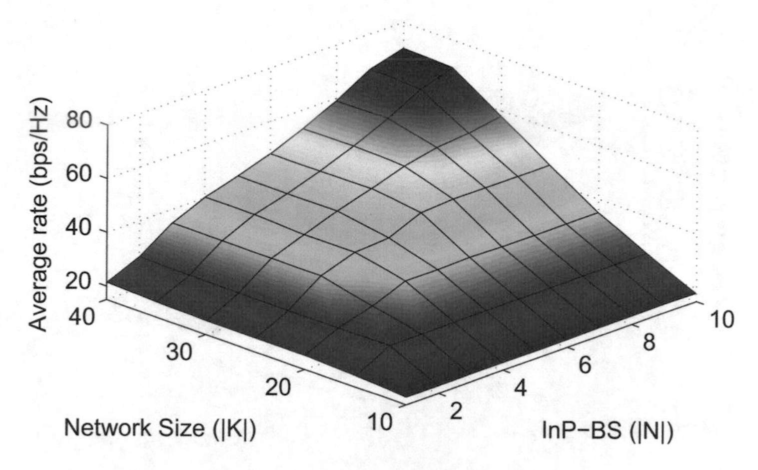

Fig. 4.8 Average sum-rate vs. network size for varying InP-BS density

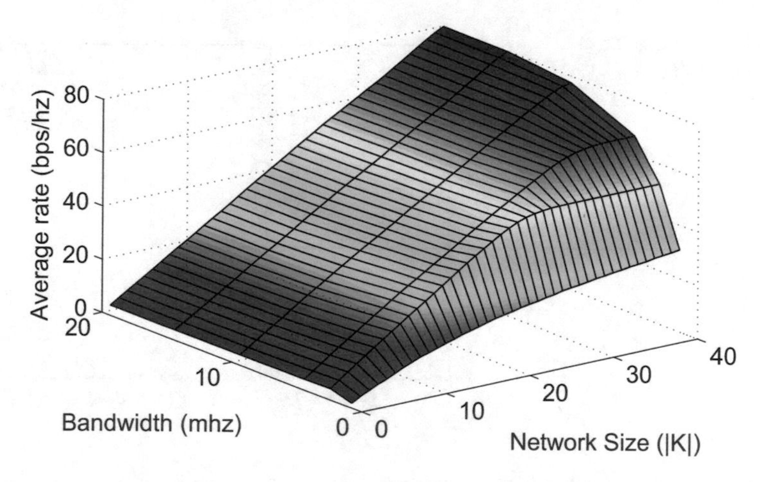

Fig. 4.9 Average sum-rate vs. network size for varying InP-BS bandwidth

up to 97% of the average sum-rate obtained by the GS scheme. Thus, it is close to optimal scheme. Furthermore, it outperforms the FS scheme by 32% for $|K| > 15$ (Figs. 4.8, 4.9, 4.10).

Next in Fig. 4.11, the slice size for each MVNO is observed. For this simulation, three InP-BS, five MVNOs, and a network size of 40 UEs were considered. In this figure, the slice size provided to each MVNO by all InP-BSs with system bandwidth 5 MHz was observed after executing the proposed algorithm. It can be seen that MVNO 2 is allocated the complete bandwidth (i.e., 25 channels) of InP-BS 3 as a slice, 80 % bandwidth from InP-BS 2 and 68 % from InP-BS 1 as a slice. This is based on its demand for each InP-BS from MVNO 2. Similarly only MVNO 1 has received the remaining 20% bandwidth of InP-BS 1 as a slice. Note that both InP-BS 1 and 3 have allocated all its channels to MVNOs whereas InP-BS

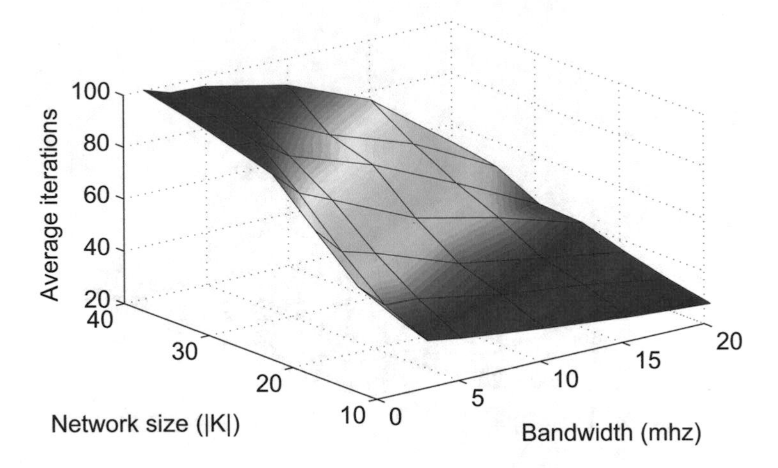

Fig. 4.10 Average iteration vs. network size for varying InP-BS bandwidth

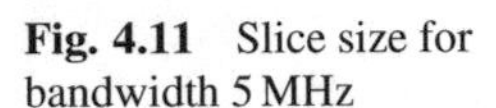

Fig. 4.11 Slice size for bandwidth 5 MHz

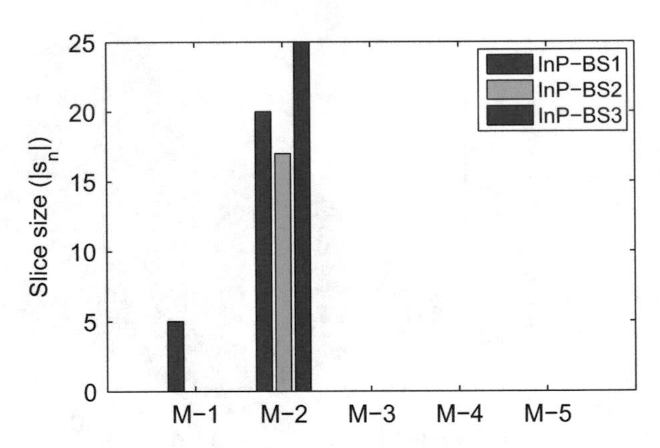

2 still have additional channels. This is due to the fact that under this scenario, no MVNO has further demand from InP-BS 2. Then, the authors rerun the simulation (considering the same simulation setup) by increasing the network bandwidth to 10 MHz. The results are shown in Fig. 4.12. Here, it was observed that MVNO 2 is allocated a larger slice (additional bandwidth) from InP-BS 3 while the slice size from other InP-BS remains the same. Note here that as the slice size from other InP-BS remains the same in both scenarios, there exist no more demands from this MVNO toward these InPs. Moreover, MVNO 1 is also allocated additional bandwidth (larger slice) from InP-BS1 and InP-BS 3. Similarly MVNO 3 is also received a share of bandwidth (slice) from InP-BS 3. However, other MVNOs are not allocated any slice. From this it can be inferred that the size of allocated slice depends upon the MVNOs demands and preference of InP-BS. For instance, in this simulation scenario, three MVNOs (MVNO 1, MVNO 2, and MVNO 3) required a slice from InP-BS 3. However, InP-BS 3 first served MVNO 2, then MVNO 1, and finally MVNO 3 following its preference profile. Finally in Fig. 4.13, we generated

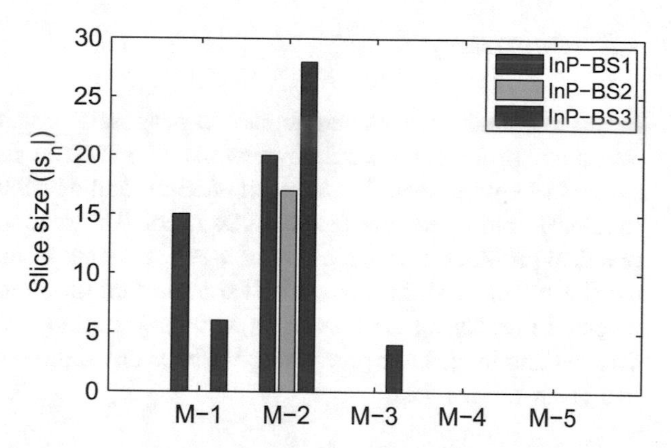

Fig. 4.12 Slice size for bandwidth 10 MHz

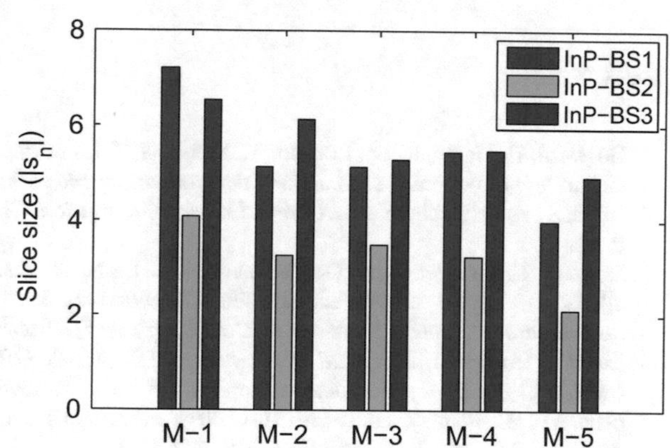

Fig. 4.13 Average size for bandwidth 10 MHz

100 random simulation instances with bandwidth 10 MHz and presented the average slice size allocated to each MVNO.

Next, in Fig. 4.8, the average sum-rate vs network size with varying number of InP-BS for the HM scheme is shown. It is inferred that as we increase the InP-BS density from 1 to 10, more opportunities are available for UEs due to increase in number of channel per serving area. Thus the average sum-rate increases with both the BS density and network size. However, when significant channels are available to fulfill all UEs demand, the average rate saturates, i.e., when $|N_j| > 9$ for a network size $|K| = 40$. The average sum-rate increases both with the network size and InP-BS density because of more resources availability in the service area.

Figure 4.9 states that the increase in average sum-rate of HM scheme when both the network size and InP-BS bandwidth increases until it gets saturated. For this simulation, the InP-BS bandwidth is increased from 1.4 to 20 MHz (i.e., 6–100 channels). Finally, the average iterations of HM scheme are shown in Fig. 4.10. It can be seen that convergence is achieved under all scenarios in a limited number of iterations.

4.3 Summary

In this chapter, we discussed about two proposals pertaining to the radio resource allocation problem in wireless network slicing. The first proposal considered a single InP environment in which an efficient and distributed scheme was designed to allocate radio resources to MVNOs users. The second proposal handles a more practical problem in which a service area consisting of multiple InPs, MVNOs, and users is considered. In this work, two important problems of service selection and resource purchasing are solved. In service selection, users choose an MVNO for services and in resource purchasing MVNOs buy radio resources from multiple InPs in a given service area.

References

1. Boccardi, F., Heath, R. W., Lozano, A., Marzetta, T. L., & Popovski, P. (2014). Five disruptive technology directions for 5G. *IEEE Communications Magazine, 52*(2), 74–80.
2. Boyd, S., & Vandenberghe, L. (2004). *Convex optimization*. Cambridge: Cambridge University Press.
3. Dawadi, R., Parsaeefard, S., Derakhshani, M., & Le-Ngoc, T. (2015). Energy-efficient resource allocation in multi-cell virtualized wireless networks. In *2015 IEEE International Conference on Ubiquitous Wireless Broadband (ICUWB)* (pp. 1–5). Piscataway: IEEE.
4. *Evolved Universal Terrestrial Radio Access (E-UTRA)*. (2011). Physical layer procedures (release 11), 650 route des lucioles. Sophia Antipolis, Valbonne, France.
5. Gale, D., & Shapley, L. S. (2013). College admissions and the stability of marriage. *The American Mathematical Monthly, 120*(5), 386–391.
6. Gu, Y., Saad, W., Bennis, M., Debbah, M., & Han, Z. (2015). Matching theory for future wireless networks: Fundamentals and applications. *IEEE Communications Magazine, 53*(5), 52–59.
7. Ho, T. M., Tran, N. H., Do, C. T., Kazmi, S. A., Huh, E.-N., & Hong, C. S. (2015). Power control for interference management and QoS guarantee in heterogeneous networks. *IEEE Communications Letters, 19*(8), 1402–1405.
8. Ho, T. M., Tran, N. H., Le, L. B., Kazmi, S. A., Moon, S. I., & Hong, C. S. (2015). Network economics approach to data offloading and resource partitioning in two-tier LTE hetnets. In *2015 IFIP/IEEE International Symposium on Integrated Network Management (IM)* (pp. 914–917). Piscataway: IEEE.
9. Kamel, M. I., Le, L. B., & Girard, A. (2014). LTE wireless network virtualization: Dynamic slicing via flexible scheduling. In *2014 IEEE 80th Vehicular Technology Conference (VTC Fall)* (pp. 1–5). Piscataway: IEEE.
10. Kamel, M. I., Le, L. B., & Girard, A. (2015). LTE multi-cell dynamic resource allocation for wireless network virtualization. In *2015 IEEE Wireless Communications and Networking Conference (WCNC)* (pp. 966–971). Piscataway: IEEE.
11. Kazmi, S. A., & Hong, C. S. (2017). A matching game approach for resource allocation in wireless network virtualization. In *Proceedings of the 11th International Conference on Ubiquitous Information Management and Communication* (p. 113). New York: ACM.
12. Kazmi, S. A., Tran, N. H., Ho, T. M., & Hong, C. S. (2018). Hierarchical matching game for service selection and resource purchasing in wireless network virtualization. *IEEE Communications Letters, 22*(1), 121–124.

13. Kazmi, S. A., Tran, N. H., Ho, T. M., Lee, D. K., & Hong, C. S. (2016). Decentralized spectrum allocation in D2D underlying cellular networks. In *2016 18th Asia-Pacific Network Operations and Management Symposium (APNOMS)* (pp. 1–6). Piscataway: IEEE.
14. Kazmi, S. A., Tran, N. H., Ho, T. M., Oo, T. Z., LeAnh, T., Moon, S., et al. (2015). Resource management in dense heterogeneous networks. In *2015 17th Asia-Pacific Network Operations and Management Symposium (APNOMS)* (pp. 440–443). Piscataway: IEEE.
15. Kazmi, S. A., Tran, N. H., & Hong, C. S. (2019). Matching games for 5G networking paradigms. In *Game theory for networking applications* (pp. 69–105). Cham: Springer.
16. Kazmi, S. A., Tran, N. H., Saad, W., Han, Z., Ho, T. M., Oo, T. Z., et al. (2017). Mode selection and resource allocation in device-to-device communications: A matching game approach. *IEEE Transactions on Mobile Computing, 16*(11), 3126–3141.
17. Kokku, R., Mahindra, R., Zhang, H., & Rangarajan, S. (2012). NVS: A substrate for virtualizing wireless resources in cellular networks. *IEEE/ACM Transactions on Networking, 20*(5), 1333–1346.
18. Kokku, R., Mahindra, R., Zhang, H., & Rangarajan, S. (2013). Cellslice: Cellular wireless resource slicing for active RAN sharing. In *2013 Fifth International Conference on Communication Systems and Networks (COMSNETS)* (pp. 1–10). Piscataway: IEEE.
19. Liang, C., & Yu, F. R. (2015). Mobile virtual network admission control and resource allocation for wireless network virtualization: A robust optimization approach. In *2015 IEEE Global Communications Conference (GLOBECOM)* (pp. 1–6). Piscataway: IEEE.
20. Liang, C., & Yu, F. R. (2015). Wireless network virtualization: A survey, some research issues and challenges. *IEEE Communications Surveys & Tutorials, 17*(1), 358–380.
21. Parsaeefard, S., Dawadi, R., Derakhshani, M., & Le-Ngoc, T. (2016). Joint user-association and resource-allocation in virtualized wireless networks. *IEEE Access, 4*, 2738–2750.
22. Roth, A. E. (2008). Deferred acceptance algorithms: History, theory, practice, and open questions. *International Journal of Game Theory, 36*(3–4), 537–569.
23. Son, K., Lee, S., Yi, Y., & Chong, S. (2011). REFIM: A practical interference management in heterogeneous wireless access networks. Preprint. arXiv:1105.0738.
24. Wen, H., Tiwary, P. K., & Le-Ngoc, T. (2013). Current trends and perspectives in wireless virtualization. In *2013 International Conference on Selected Topics in Mobile and Wireless Networking (MoWNeT)* (pp. 62–67). Piscataway: IEEE.
25. Zhang, G., Yang, K., Wei, J., Xu, K., & Liu, P. (2015). Virtual resource allocation for wireless virtualization networks using market equilibrium theory. In *2015 IEEE Conference on Computer Communications Workshops (INFOCOM WKSHPS)* (pp. 366–371). Piscataway: IEEE.
26. Zhu, K., & Hossain, E. (2016). Virtualization of 5G cellular networks as a hierarchical combinatorial auction. *IEEE Transactions on Mobile Computing, 15*(10), 2640–2654.

Chapter 5
Network Slicing: Radio Resource Allocation Using Non-orthogonal Multiple Access

5.1 Introduction

Virtualization of a network is considered as one of the essential formative viewpoints in the pushing toward fifth generation (5G) cellular networks. Many inherent benefits such as enhanced data rates, spectrum/energy efficiency, capacity, and lower latency are expected to be achieved for higher quality of experience [11, 12, 16].

However, the unprecedented growth in data traffic and the tsunami of mobile devices in the existing networks demands to consider spectrum efficiency and massive connectivity in 5G networks. Also, the greatest limitation of the existing employed access scheme, i.e., orthogonal multiple access (OMA) scheme (i.e., the OFDMA scheme) is that the number of users that can simultaneously occupy the spectrum resources is restricted by the number of available spectrum resources, i.e., subchannels, resource blocks (RBs). Recently, a new promising technology has been under consideration for solving the challenges of OMA-based schemes. Non-orthogonal multiple access (NOMA) has been viewed as a key enabler for catering the inconveniences of OFDMA scheme in 5G networks [14, 15]. NOMA takes advantage of the resource gain differences through which it allows multiple users to be scheduled on a single spectrum resource. Thus, the NOMA technique can accommodate higher number of users in the network when compared to traditional OMA schemes. However, users that are scheduled over the same spectrum resource create inter-user interference over the spectrum resource. This problem can be solved by using the successive interference cancellation (SIC) technique at the receiver. However, reaping the benefits of NOMA requires meeting significant challenges in terms of resource and power assignment [7, 9].

S. M. A. Kazmi et al., *Network Slicing for 5G and Beyond Networks*,
https://doi.org/10.1007/978-3-030-16170-5_5

5.2 System Model and Problem Formulation

Consider the downlink of the system model illustrated in Fig. 5.1, which consists of a single cell with a macro base station (MBS) [7]. The third party service provider such as infrastructure provider (InP) owns and manages the spectrum and MBS. The virtual network services using individual contracts are provided by InP to slices set $\mathcal{N}$. The set $\mathcal{U}_n$ of users are provided with services by a slice $n \in \mathcal{N}$. Let the set of all users is denoted by $\mathcal{U} \triangleq \cup_{n=1}^{|\mathcal{N}|}\mathcal{U}_n$. Further, the system bandwidth $\mathcal{C}$ that consists of Ω RBs is owned by the InP.

In model of [7], the NOMA cluster is formed by a set of users that are scheduled for a non-orthogonal set of same resource blocks. The set of resource blocks on which the NOMA cluster operates is orthogonal to other NOMA clusters resource blocks. Along with this, the number of users ranges from 2 to $|\mathcal{U}|$ in a NOMA cluster. Apart from this, δ^k (where $1 \leq \delta^k \leq \Omega$) represents the k-th cluster allocated resource blocks. Let the clusters set, the active users set that are grouped into k-th cluster, and maximum MBS transmission power budge are represented by $\mathcal{S}$, $\mathcal{S}_k$, and P_T respectively. P_{j_n} and $h_{j_n} = \chi_{j_n}/\mathcal{D}(d_{j_n})$ denote the allocated power to user j_n and complex channel coefficient between MBS and user j_n. Where χ_{j_n}, $\mathcal{D}(d_{j_n})$, and d_{j_n} represent the gain of the Rayleigh fading channel, path-loss function, and distance between the MBS and the user j_n respectively. Let the symbol of user j_n

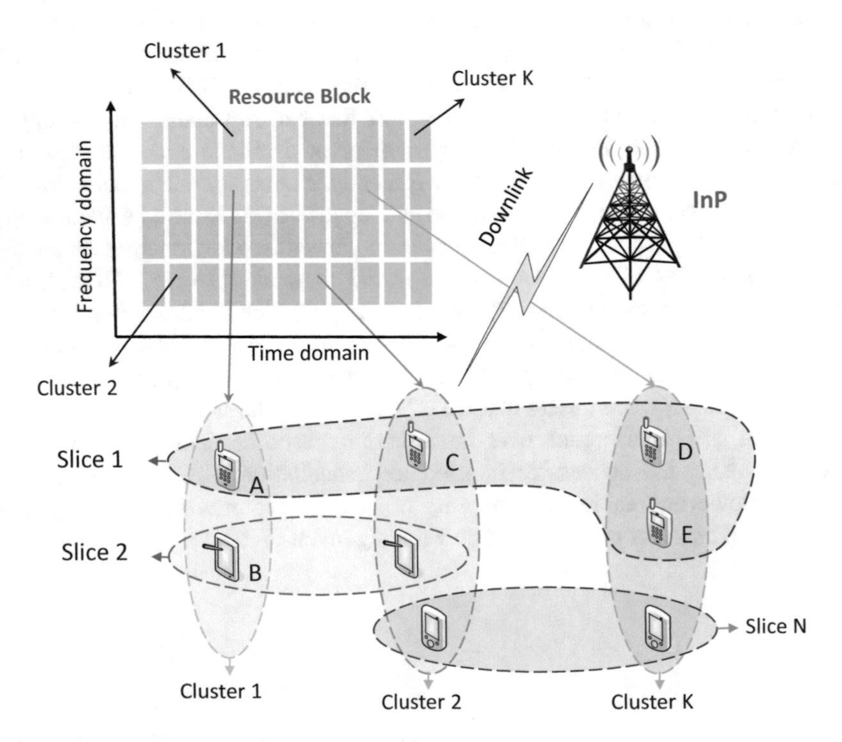

Fig. 5.1 System model of WNV in which users of different MVNOs coexist on a group of resource blocks owned by an InPs using NOMA

that is transmitted is represented by x_{j_n}. In a k-th cluster, the MBS signal that is received by user j_n is given by:

$$y^k_{j_n} = h_{j_n}\sqrt{P_{j_n}}x^k_{j_n} + \sum_{i \neq j | i \in \mathcal{S}_k} h_{j_n}\sqrt{P_{i_m}}x^k_{i_m} + z_{j_n}, \tag{5.1}$$

where z_{j_n} denotes the noise. Within a cluster, all its users can have access to the resource blocks which might result in interference among the users. Every user will use SIC on the received superimposed signals [5]. Generally, lower power levels are assigned to users with high channel gain. The signals with lower channel power are recovered in SIC decoding later than the signals with high power levels. In a similar fashion, the signals of the users with lower power levels are recovered through treatment of the lower power level signals of other users as noise in SIC decoding [1, 5].

Optimally, the decoding order of the SIC should follow increasing channel gains normalized by noise. More specifically, the user $j_n \in \mathcal{S}_k$ receiver can have the ability to cancel interference of other user $i_m \in \mathcal{S}_k$ having channel gain $|h_{i_m}|^2/z_{i_m} < |h_{j_n}|^2/z_{j_n}$. The clustering variable $\beta^k_{j_n}$ is defined as follows:

$$\beta^k_{j_n} = \begin{cases} 1, & \text{if user } j_n \text{ is grouped into cluster } k, \\ 0, & \text{otherwise.} \end{cases}$$

Thus, the data rate for a user j_n in the downlink NOMA k-th cluster is given as follows:

$$R^k_{j_n} = \delta^k B \log_2\left(1 + \frac{P_{j_n}|h_{j_n}|^2}{I^k_{j_n} + \delta^k B z_{j_n}}\right), \tag{5.2}$$

where $I^k_{j_n}$ denote the interference perceived at user $j_n \in \mathcal{U}_n$ due to coexisting other users on the same k-th cluster.

$$I^k_{j_n} = \sum\nolimits_{i_m \in \mathcal{U} | \frac{|h_{i_m}|^2}{z_{i_m}} > \frac{|h_{j_n}|^2}{z_{j_n}}} \beta^k_{i_m} P_{i_m}|h_{j_n}|^2. \tag{5.3}$$

In NOMA, the users with higher channel gains are the only source of interference. For example, consider a cluster of two users that follow $\frac{|h_1|^2}{z_1} > \frac{|h_2|^2}{z_2}$. The user 1 first performs the decoding of x_2, then performs its subtraction from the signal y^k, and then finally performs the decoding of the signal x_1. Other than that, the user 2 will not do interference cancellation. Thus, we have:

$$R^k_1 = \delta^k B \log_2\left(1 + \frac{P_1|h_1|^2}{\delta^k B z_1}\right),$$

$$R^k_2 = \delta^k B \log_2\left(1 + \frac{P_2|h_2|^2}{P_1|h_2|^2 + \delta^k B z_2}\right).$$

Obtaining a better resource allocation along with isolation in WNV is one of the major challenges in its implementation. WNV enables the InP to create slices of physical resources. These slices are isolated which make it possible to make changes in any one of the slices without affecting the others [11]. At the physical resource level isolation, implementation of the isolation can be performed using different techniques. One approach is to assign each slice a fixed amount of predefined physical resources while in other cases, dynamic sharing is done and no restriction on the resource access is made. However, in dynamic sharing, the isolation is achieved using contract agreements [17]. This work considered the dynamic sharing scheme that is based on guaranteeing the service contract agreements between the InP and MVNOs by a minimum data rate [4, 17]. Let the minimum data rate of the n-th slice is represented by $R_n^{\min}$. Then the isolation constraint that must be fulfilled in creating NOMA clusters by InP is given by:

$$\sum_{j_n \in \mathcal{U}_n} \sum_{k \in \mathcal{S}} \beta_{j_n}^k R_{j_n}^k \geq R_n^{\min}, \ \forall n \in \mathcal{N}. \tag{5.4}$$

5.2.1 Problem Formulation

The authors in [7] considered the network sum-rate objective and a weight factor ω_{j_n} is used for user j_n to enable adjustment of the priorities of the users. In the downlink NOMA, the joint problem of resource block allocation, power assignment problem, and user clustering for weighted sum-rate maximization can be formulated as:

$$\begin{aligned}
&\max_{\beta,\delta,\mathbf{P}} \sum_{n \in \mathcal{N}} \sum_{j_n \in \mathcal{U}_n} \omega_{j_n} \sum_{k \in \mathcal{S}} \beta_{j_n}^k \delta^k B \log_2 \left(1 + \frac{P_{j_n} |h_{j_n}|^2}{I_{j_n}^k + \delta^k B z_{j_n}}\right) \\
&\text{s.t.:} \ C_1 : \sum_{n \in \mathcal{N}} \sum_{j_n \in \mathcal{U}_n} \sum_{k \in \mathcal{S}} \beta_{j_n}^k P_{j_n} \leq P_T, \\
&C_2 : \sum_{j_n \in \mathcal{U}_n} \sum_{k \in \mathcal{S}} \beta_{j_n}^k \delta^k B \log_2 \left(1 + \frac{P_{j_n} |h_{j_n}|^2}{I_{j_n}^k + \delta^k B z_{j_n}}\right) \geq R_n^{\min}, \ \forall n \\
&C_3 : \sum_{k \in \mathcal{S}} \beta_{j_n}^k = 1, \ \forall j_n \in \mathcal{U}_n, n \in \mathcal{N}, \\
&C_4 : 2 \leq \sum_{n \in \mathcal{N}} \sum_{j_n \in \mathcal{U}_n} \beta_{j_n}^k \leq |\mathcal{U}|, \ \forall k, \\
&C_5 : \sum_{k \in \mathcal{S}} \beta_{j_n}^k \delta^k \leq \Omega, \ \forall j, n, \\
&C_6 : \delta^k \in \{1, 2, \ldots, \Omega\}, \ \beta_{j_n}^k \in \{0, 1\}, \ \forall k, j, n,
\end{aligned} \tag{5.5}$$

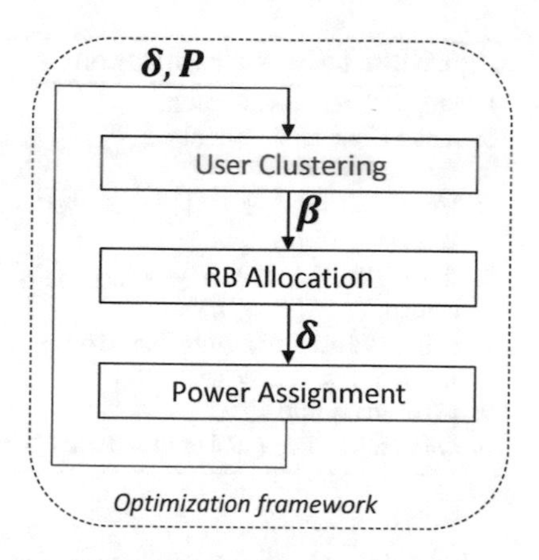

Fig. 5.2 Proposed framework [7]

where the power assignment vector is given by $\mathbf{P} \triangleq \{P_{j_n}\}$, $\forall j_n \in \mathcal{U}$, the RB allocation vector is represented by $\boldsymbol{\delta} \triangleq \{\delta^k\}$, $\forall k \in \mathcal{S}$, and the user clustering matrix is represented by $\boldsymbol{\beta} \triangleq \{\beta^k_{j_n}\}$, $\forall j_n \in \mathcal{U}, k \in \mathcal{S}$, respectively. The constraints C_1, C_2, C_3, and C_4 are used to represent the total power constraint, ensure slice isolation requirement in NOMA clustering, ensure assignment of a user to at most a single cluster, and to ensure that at least two users are grouped into NOMA cluster. The constraint C_5 denotes the spectrum resource constraint on a downlink.

The problem of (5.5) is formulated as a mixed-integer non-convex optimization problem. For tackling the above problem, the decoupling of resource allocation and user clustering has been performed. Then a joint solution that involves the iterative solution of resource allocation and user clustering is proposed as shown in Fig. 5.2.

5.3 Solution Approach

5.3.1 *Matching Game for User Clustering (NOMA Clustering)*

The working of NOMA is assisted by users grouping with different channel gains over a resource block. The users are divided into three categories according to channel gains, i.e., strong users, normal users, and weak users.

Users Classification

First of all, the classification of the users has been performed in this section and then total clusters in a network are computed. The reason for using this intuition is due to the fact that a single NOMA cluster has users with different channel gains. The scheme used for clustering and classification has two steps. The initial step is the classification phase that is done through the sorting of users in descending

Algorithm 1 Users classification

1: **Step 1**: User classification:
2: Input: Channel threshold: θ_A, θ_B,
3: Set $A, B, C = \emptyset$.
4: Sort users: $|h_1|^2 \geq |h_1|^2 \cdots \geq |h_i|^2 \geq \theta_A > |h_{i+1}|^2 \geq |h_{i+2}|^2 \geq \cdots \geq |h_j|^2 \geq \theta_B > |h_{j+1}|^2 \geq \cdots |h_{|\mathcal{U}|}|^2$.
5: $A \leftarrow \{i||h_i|^2 \geq \theta_A\}$; $B \leftarrow \{j|\theta_A \geq |h_j|^2 \geq \theta_B\}$; $C \leftarrow \mathcal{U}\backslash\{A, B\}$
6: Output: $\mathcal{K} = \{A, B, C\}$
7: **Step 2**: Cluster formation based on classification:
8: Input: $|\mathcal{S}| = 0$, $\mathcal{K} = \{A, B, C\}$
9: $|\mathcal{S}| = \max_i |\mathcal{K}_i|$
10: Output: $|\mathcal{S}|$, i.e., number of required NOMA clusters

order (Line 4). Following users classification, the clustering of users into three different groups is performed. The next step 2 is the computation of the number of required clusters utilizing the output of the previous step. The number of clusters $\mathcal{S}$ is taken equal to maximum cardinality after computing the cardinality of all classes. Following the computation of the clusters and users classes, the next step is the grouping of users into clusters.

This subproblem can be solved by assuming that power level assignment of every user and number of resource blocks given to each cluster are given. The users clustering problem is as follows:

$$\begin{aligned}
\mathbf{UC}: \max_{\beta} & \sum_{n\in\mathcal{N}} \sum_{j_n\in\mathcal{U}_n} \omega_{j_n} \sum_{k\in\mathcal{S}} \beta_{j_n}^k R_{j_n}^k \\
\text{s.t.:}\ & C_1: \sum_{n\in\mathcal{N}} \sum_{j_n\in\mathcal{U}_n} \sum_{k\in\mathcal{S}} \beta_{j_n}^k P_{j_n} \leq P_T, \\
& C_2: \sum_{j_n\in\mathcal{U}_n} \sum_{k\in\mathcal{S}} \beta_{j_n}^k R_{j_n}^k \geq R_n^{\min},\ \forall n \in \mathcal{N}, \\
& C_3: \sum_{k\in\mathcal{S}} \beta_{j_n}^k = 1,\ \forall j_n \in \mathcal{U}_n, n \in \mathcal{N}, \\
& C_4: 2 \leq \sum_{n\in\mathcal{N}} \sum_{j_n\in\mathcal{U}_n} \beta_{j_n}^k \leq |\mathcal{U}|,\ \forall k \in \mathcal{S}, \\
& C_5: \sum_{k\in\mathcal{S}} \beta_{j_n}^k \delta^k \leq \Omega,\ \forall j_n \in \mathcal{U}_n, n \in \mathcal{N}, \\
& C_6: \beta_{j_n}^k \in \{0, 1\},\ \forall k, j, n.
\end{aligned} \tag{5.6}$$

Finding the solution of the problem in (5.6) is NP-hard for a large number of clusters and users due to its combinatorial nature [10]. Therefore it's desirable to find the solution of the problem (5.6) in a distributed manner. Problem (5.6) is solved using a matching game that does not consider the C_1 constraint. Similarly the constraint C_4 is also relaxed through allowing a single user in a cluster.

Matching Game with Externalities for User Clustering

The execution of cluster formation and user classification results in the user classes and the required number of clusters at iteration t. Then, for a given number of clusters, a user clustering is carried out. This section uses a matching theory for performing clustering of users. The reason for using matching game is because of its ability to solve problems that have combinatorial nature [6, 8, 10]. The distributed nature of the matching game is exploited in WNV and NOMA to enable complexity reduction of centralized approach on InP. Additionally, matching theory enables players to perform utility definition through its local information. In this model, the sets of agents are the MVNO users defined by the set, $\mathcal{U}$, and the clusters defined by the set, $\mathcal{S}$. The preference profile $\mathcal{P}_k$ of the kth cluster is complete, transitive, and strict. From constraint C_3 in (5.6), it is clear that a user can be associated with one cluster. On the other hand, different users can be associated with a single cluster. Therefore, user preference profile $\mathcal{P}_u$ has been defined over the clusters $\mathcal{S}$. As the preference ranking of a user j is affected by other users in the same cluster, therefore, one-to-many matching given by tuple $(\mathcal{U}, \mathcal{S}, \succ_{\mathcal{U}}, \succ_{\mathcal{S}})$ is considered. The matching is defined as follows.

Definition 5.1 A *matching* β is defined as a function from the set $\mathcal{U} \cup \mathcal{S}$ into the set $\mathcal{U} \cup \mathcal{S}$ which satisfies for all $k \in \mathcal{S}$ and $j \in \mathcal{U}$ [7]:

1. $|\beta(j)| \leq 1$ and $\beta(j) \in \mathcal{S} \cup \phi$,
2. $|\beta(k)| \leq q_k$ and $\beta(k) \in 2^{\mathcal{U}} \cup \phi$,
3. If $j \in \beta(k)$ then $\beta(j) = k$,
4. If $\beta(j) \in k$ for cluster k then $\beta(k) = j$,

where q_k and $|\beta(.)|$ represent the kth cluster quota and matching outcome $\beta(.)$ cardinality, respectively. In Definition 7.1, the first two conditions denote C_4 and C_3 in (5.6). Where $q_k \leq |\mathcal{U}|$ denote the kth cluster total quota. The $\beta(j) = \phi$ gives indication that j has not been matched to any cluster. Similarly, $\beta(k) = \phi$ means that no user is matched to the kth cluster.

Both sides require to perform ranking of each other in the formulated game of [7] through preference profiles. On the other side, the users preference profilers are affected by offered resources to a cluster along with other users in that cluster. This type of interdependences in matching theory is called as *externalities* which has significant impact on the design of the proposed solution. An agent might change its preference order continuously because of these externalities in response to other agents activities. Therefore, care should be taken during the design of solution based on matching game to attain convergence. Each user computes data rate that can be achieved for every cluster and then place them in descending order to build preference profile $(\mathcal{P}_j, \ \forall j \in \mathcal{U})$. Moreover, the minimum data rate requirement over a slice has been transformed by relaxing to a user minimum data rate, i.e., $R_j^{\min} = R_n^{\min} \frac{|h_{jn}|^2}{\sum_{jn \in \mathcal{U}_n} |h_{jn}|^2}$, $\forall j \in \mathcal{U}_n$, to enable matching game. Therefore, utility of a user can be written as:

$$U_j(k, \beta) = \max(R_j^k, R_j^{\min}), \quad \forall k \in \mathcal{S}. \tag{5.7}$$

Thus, the preference relation $\succ_j$ for any user j is defined over the set of clusters $\mathcal{S}$ which can be given as follows:

$$(k, \beta) \succ_j (k', \beta') \Leftrightarrow U_j(k, \beta) > U_j(k', \beta'). \tag{5.8}$$

Similarly, each cluster k aims to choose a set of users that can maximize its utility. Therefore, the cluster uses the following utility to create its preference profile $(\mathcal{P}_k)$

$$U_k(j, \beta) = \max(R_j^s - R_j^{\min}, 0), \quad \forall j \in \mathcal{U}, \tag{5.9}$$

Each cluster k as per (5.9) performs user selection whose data rate requirements of minimum data rate can be fulfilled and to maximize the cluster data rate. The user is ranked least and denoted by zero preference in case if the demand of the user cannot be met, i.e., constraint C_5. A preference relation $\succ_k$ for a cluster k is defined as follows:

$$(j, \beta) \succ_s (j', \beta') \Leftrightarrow U_k(j, \beta) > U_k(j', \beta'). \tag{5.10}$$

After defining preference profiles along with the matching game, a stable clustering for a game is defined while taking into the account the externalities effect. It is clear from (5.9) and (5.7) that preferences are a function of the current matching β and interference of higher SINR users that affects other users performance. To tackle such type of externalities, a scheme is discussed in the next subsection.

In the proposed game, any user j grouped with users of same cluster k will induce interference on other users that are grouped in the same cluster in case its gain is higher than other users of the same cluster. Apart from that, a user might change its preference order due to action of other users that reside in the same cluster. The situation might produce a case where the final clustering will never be attained. To tackle this issue, the proposed algorithm uses the broadcast of initial network information by InP. It is assumed that users will sent such type of information to InP during the initialization step. Each user can compute the users set that have higher channel gains using this type of information. The users that are of the same class are only taken into consideration. The externality set by $\mathcal{C}_j$ is as follows:

$$\mathcal{C}_j = \left\{ j' \in \mathcal{U} : \frac{|h_{j'}|^2}{z_{j'}} > \frac{|h_j|^2}{z_j}, j, j' \in \mathcal{A} \right\}, \tag{5.11}$$

Algorithm 2 User clustering algorithm

1: **input**: $\mathcal{P}_j^{(t)}, \mathcal{P}_k^{(t)}, \mathcal{C}_j, \forall k, j$.
2: **initialize**: $t = 0$, $\beta^{(1)} \triangleq \{\beta(j)^{(1)}, \beta(k)^{(1)}\}_{j\in\mathcal{U},k\in\mathcal{S}} = \emptyset$, $\mathcal{J}_k{}^{(1)} = \emptyset$, $C_s^{(1)} = \emptyset$, $q_k^{(1)} = |\mathcal{U}|$, $\forall k, j$.
3: **repeat**
4: $t \leftarrow t + 1$.
5: Update $\forall j, \mathcal{P}_j{}^{(t)}$ for given $\beta(s)^{(t-1)}$.
6: $\forall j \in$ class $\mathcal{K}$ with k as its most preferred in $\mathcal{P}_j^{(t)}$.
7: **while** $j \notin \beta(k)^{(t)}$ and $\mathcal{P}_j^{(t)} \neq \emptyset$ **do**
8: **if** $C_k^{(t)} = \{j' \in \beta(k)^{(t)} \cup \mathcal{C}_j\} \neq \emptyset$ **then**
9: $\mathcal{X'}_k^{(t)} = \{j' \in \beta(k)^{(t)}, k' \in \mathcal{C}_j | j \succ_k j'\}$.
10: $j_{lp} \leftarrow$ the least preferred $j' \in \mathcal{X'}_k^{(t)}$.
11: **for** $j_{lp} \in \mathcal{X'}_k^{(t)}$ **do**
12: $\beta(k)^{(t)} \leftarrow \beta(k)^{(t)} \setminus j_{lp}, q_k{}^{(t)} \leftarrow q_k{}^{(t)} + 1$.
13: **end for**
14: **if** $C_k^{(t)} = \{j' \in \beta(k)^{(t)} \cup \mathcal{C}_j\} \neq \emptyset$ **then**
15: $j_{lp} \leftarrow j$.
16: **else**
17: $\beta(k)^{(t)} \leftarrow \beta(s)^{(t)} \cup j, q_k{}^{(t)} \leftarrow q_k{}^{(t)} - 1$.
18: **end if**
19: **else**
20: **if** Check $q_k{}^{(t)} > 0$ **then**
21: $\beta(k)^{(t)} \leftarrow \beta(k)^{(t)} \cup j, q_k{}^{(t)} \leftarrow q_k{}^{(t)} - 1$.
22: **else**
23: $j_{lp} \leftarrow j$.
24: **end if**
25: **end if**
26: $\mathcal{J}_k{}^{(t)} = \{j \in \mathcal{X'}_k{}^{(t)} | j_{lp} \succ_k j\} \cup \{k_{lp}\}$.
27: **for** $j \in \mathcal{J}_r{}^{(t)}$ **do**
28: $\mathcal{P}_j{}^{(t)} \leftarrow \mathcal{P}_j{}^{(t)} \setminus k\ \mathcal{P}_k{}^{(t)} \leftarrow \mathcal{P}_k{}^{(t)} \setminus j$.
29: **end for**
30: **end while**
31: **Check**: $\beta^{(t-1)} = \beta^{(t)}$.
32: **until** $\forall$ classes, i.e., A, B, C.

where $\mathcal{A}$ denote same class from Algorithm 1. The users of the same class that have higher gain than user j is selected using (5.11). The purpose of this type of externality set is to enable restriction on users of same class to do grouping within same cluster.

Clustering Algorithm

To determine a stable clustering, the initial step is to define the game blocking pair. The Gale–Shapley style algorithm does not apply to the game because of the externalities in proposed matching game. Therefore, the blocking pair is designed for game through the algorithm as follows.

Definition 5.2 A matching β is blocked by a pair of players (j, k) if there exists a pair (j, k) such that, $j \succ_k \beta(k)$, $k \succ_j \beta(j)$, and $\beta(k) \notin \mathcal{C}_j$.

The intuition used in Definition 5.2 is: Whenever a cluster k is preferred by a user j compared to its assigned cluster $\beta(j)$ which has no user conflicts (i.e., $\beta(k) \notin \mathcal{C}_j$). Apart from that the cluster k preferred to accept j, then a blocking pair can be formed by cluster k and user j through deviation from assigned matching. The existence of stable matching is characterized by the absence of blocking pair. Moreover, a condition for attaining the stability is the creation of newly generated pair of agents must not undermine the existing pairs stability. Using this condition will allow the preference profiles of already matched pairs of agents to remain unchanged due to the creation of new pair of agents.

Algorithm 2 presents the pseudocode of the proposed user clustering scheme proposed by Ho et al. [7]. The detailed working of this scheme has been described in [7]. Moreover, the authors have also proved that this scheme converges to a stable solution that has no blocking pairs. Formally, it is stated as follows.

Theorem 5.1 *A convergence to stable allocation is achieved by Algorithm 2 [7].*

Proof The proof is presented in Theorem 1 of [7].

Once the user clustering problem is solved, the aim is to find the appropriate resources required to meet the requirements of each cluster. Therefore, next, the authors solve the resource allocation problem for each cluster.

5.3.2 Resource Allocation

The problem of resource blocks allocation to every cluster, given power level assignment $\mathbf{P}$ and user clustering $\boldsymbol{\beta}$, is formulated as follows:

$$\mathbf{RBA}: \max_{\delta} \sum_{k \in \mathcal{S}} \sum_{j_n \in \mathcal{S}_k} \omega_{j_n} \delta^k B \log_2 \left(1 + \frac{P_{j_n} |h_{j_n}|^2}{I^k_{j_n} + \delta^k B z_{j_n}}\right)$$

s.t.:

$$C_2: \sum_{j_n \in \mathcal{U}_n} \sum_{k \in \mathcal{S}} \beta^k_{j_n} \delta^k B \log_2 \left(1 + \frac{P_{j_n} |h_{j_n}|^2}{I^k_{j_n} + \delta^k B z_{j_n}}\right) \geq R^{\min}_n, \ \forall n \quad (5.12)$$

$$C_5: \sum_{k \in \mathcal{S}} \delta^k \leq \Omega,$$

$$C_6: \delta^k \in \{1, 2, \ldots, \Omega\}, \ \forall k,$$

where $I^k_{j_n} = \sum_{i_m \in \mathcal{S}_k | \frac{|h_{i_m}|^2}{z_{i_m}} > \frac{|h_{j_n}|^2}{z_{j_n}}} P_{i_m} |h_{j_n}|^2$.

There is no polynomial algorithm to solve problem (5.12) because of its integer programming nature. The problem is simplified through relaxing discrete variables δ^k into real numbers in the interval $[1, \Omega]$. The simplified problem is now a concave problem due to the fact that constraints C_2 and objective have the form of $x\log(1+\frac{1}{x})$.

Proposition 5.1 *Problem* (5.12) *objective function has concave nature having positive δ^k. Therefore the optimal resource block allocation for every cluster k satisfies:*

$$\begin{aligned}\frac{\partial L(\delta^k, \lambda_{j_n})}{\partial \delta^k} &= \sum_{j_n \in \mathcal{S}_k} \omega_{j_n}\left(\mathcal{R}^k_{j_n} - \frac{\delta^{k*}\mathcal{H}^k_{j_n}}{\mathcal{I}^k_{j_n}\mathcal{Q}^k_{j_n}}\right) \\ &+ \sum_{n\in\mathcal{M}} \lambda_n \sum_{j_n\in\mathcal{U}_n} \beta^k_{j_n}\left(\mathcal{R}^k_{j_n} - \frac{\delta^{k*}\mathcal{H}^k_{j_n}}{\mathcal{I}^k_{j_n}\mathcal{Q}^k_{j_n}}\right) = 0,\end{aligned} \tag{5.13}$$

where λ_n and $L(\delta^k, \lambda_{j_n})$ are the multiplier corresponding to the isolation constraints C_2 and Lagrangian, respectively.

$$\begin{aligned}\mathcal{I}^k_{j_n} &= I^k_{j_n} + \delta^{k*}Bz_{j_n}, \\ \mathcal{Q}^k_{j_n} &= I^k_{j_n} + P_{j_n}|h_{j_n}|^2 + \delta^{k*}Bz_{j_n}, \\ \mathcal{H}^k_{j_n} &= P_{j_n}|h_{j_n}|^2B^2z_{j_n}, \\ \mathcal{R}^k_{j_n} &= B\log_2\left(1+\frac{P_{j_n}|h_{j_n}|^2}{I^k_{j_n}+\delta^{k*}Bz_{j_n}}\right).\end{aligned} \tag{5.14}$$

The solution of (5.12) is obtained using dual-based approach given in Algorithm 3.

Theorem 5.2 *Algorithm 3 is used to find the sub-optimal solution of* (5.12).

The common scheme used to solve the problem considering the context of resource blocks allocation of LTE involves the mapping of the continuous solutions in Algorithm 3 to the largest previous integer, i.e., $\overline{\delta}^* = \lfloor\delta^*\rfloor$, where $\lfloor x\rfloor$ denotes the largest integer not more than x.

5.3.3 Power Assignment

In this section, the non-convex problem of power assignment is presented. Then the algorithm using the arithmetic–geometric mean approximation for efficient power assignment is proposed [3, 13]. Finally, a scheme for joint power assignment,

Algorithm 3 Dual-based resource allocation

1: **repeat**

2: Update the number of RB $\delta^k[t+1]$:

$$\delta^k[t+1] = \frac{\sum_{j_n \in \mathcal{S}_k} \omega_{j_n} \mathcal{R}_{j_n}^k[t] + \sum_{n \in \mathcal{M}} \lambda_n \sum_{j_n \in \mathcal{U}_n} \beta_{j_n}^k \mathcal{R}_{j_n}^k[t]}{\sum_{j_n \in \mathcal{S}_k} \frac{\mathcal{H}_{j_n}^k}{\mathcal{I}_{j_n}^k[t]\mathcal{Q}_{j_n}^k[t]} + \sum_{n \in \mathcal{M}} \lambda_n \sum_{j_n \in \mathcal{U}_n} \beta_{j_n}^k \frac{\mathcal{H}_{j_n}^k}{\mathcal{I}_{j_n}^k[t]\mathcal{Q}_{j_n}^k[t]}} \tag{5.15}$$

$$\lambda_n[t+1] = \left[\lambda_n[t] - \tau[t]\left(\sum_{k \in \mathcal{S}} \sum_{j_n \in \mathcal{U}_n} \beta_{j_n}^k \delta^k[t] \mathcal{R}_{j_n}^k[t] - R_n^{\min}\right)\right]^+ \tag{5.16}$$

3: Normalize $\delta^k[t+1]$:

$$\delta^k[t+1] \leftarrow \min\left\{\delta^k[t+1], \frac{\delta^k[t+1]\Omega}{\sum_{s \in \mathcal{S}} \delta^s[t+1]}\right\}, \tag{5.17}$$

4: **until** Converge;

resource block allocation, and user clustering is proposed. The power assignment problem using resource allocation $\boldsymbol{\delta}$ and user clustering $\boldsymbol{\beta}$ is formulated as:

$$\begin{aligned}
\mathbf{PA1}: \ & \max_{\mathbf{P}} \sum_{k \in \mathcal{S}} \sum_{j_n \in \mathcal{S}_k} \omega_{j_n} \beta_{j_n}^k \delta^k B \log_2\left(1 + \frac{P_{j_n}|h_{j_n}|^2}{I_{j_n}^k + \delta^k B z_{j_n}}\right) \\
\text{s.t.:} \ & C_1: \sum_{k \in \mathcal{S}} \sum_{j_n \in \mathcal{S}_k} P_{j_n} \le P_T, \\
& C_2: \sum_{j_n \in \mathcal{U}_n} \sum_{k \in \mathcal{S}} \beta_{j_n}^k \delta^k B \log_2\left(1 + \frac{P_{j_n}|h_{j_n}|^2}{I_{j_n}^k + \delta^k B z_{j_n}}\right) \ge R_n^{\min}, \ \forall n.
\end{aligned} \tag{5.18}$$

This problem apparently seems non-convex due to the non-concave nature of the rate function in (5.2). To tackle this difficulty, a successive convex approximation (SCA) is adopted to compute the optimal power assignment.

Arithmetic–Geometric Mean Approximation

Let's define a subset

$$\begin{aligned}
\bar{\mathcal{S}}_k(j_n) &\triangleq \left\{ i_m \in \mathcal{S}_k | \frac{|h_{i_m}|^2}{z_{i_m}} > \frac{|h_{j_n}|^2}{z_{j_n}} \right\}, \\
\tilde{\beta}^k_{j_n} &\triangleq \beta^k_{j_n} \delta^k B, \\
\tilde{z}_{j_n} &\triangleq z_{j_n} \delta^k B,
\end{aligned} \tag{5.19}$$

of user throughput $j_n \in \mathcal{S}_k$ which can be rewritten as:

$$R^k_{j_n} = \log_2 \left(\frac{\sum\limits_{i_m \in \bar{\mathcal{S}}_k(j_n)} P_{i_m} |h_{j_n}|^2 + \tilde{z}_{j_n} + P_{j_n} |h_{j_n}|^2}{\sum\limits_{i_m \in \bar{\mathcal{S}}_k(j_n)} P_{i_m} |h_{j_n}|^2 + \tilde{z}_{j_n}} \right)^{\delta^k B}. \tag{5.20}$$

The problem (5.18) can be shown equivalent to:

Algorithm 4 SCA-based power assignment with arithmetic–geometric mean approximation

---

1: Initialize: $t = 1$;
2: **repeat**
3: Compute each coefficient

$$\kappa_{i_m}[t] = \frac{P_{i_m}[t-1]|h_{j_n}|^2}{u^k_{j_n}(\mathbf{P}[t-1])}, \lambda_{j_n}[t] = \frac{\tilde{z}_{j_n}}{u^k_{j_n}(\mathbf{P}[t-1])},$$

$$\gamma_{j_n}[t] = \frac{P_{j_n}[t-1]|h_{j_n}|^2}{u^k_{j_n}(\mathbf{P}[t-1])}.$$

4: Compute monomial

$$\begin{aligned}
\underline{u}^k_{j_n}(\mathbf{P})[t] = &\prod_{i_m \in \bar{\mathcal{S}}_k(j_n)} \left(\frac{P_{i_m}[t-1]|h_{j_n}|^2}{\kappa_{i_m}[t]} \right)^{\kappa_{i_m}[t]} \\
&\times \left(\frac{\tilde{z}_{j_n}}{\lambda_{j_n}[t]} \right)^{\lambda_{j_n}[t]} \left(\frac{P_{j_n}[t-1]|h_{j_n}|^2}{\gamma_{j_n}[t]} \right)^{\gamma_{j_n}[t]}.
\end{aligned} \tag{5.21}$$

5: With $\underline{u}^k_{j_n}(\mathbf{P})[t]$, solve the geometric program (5.24), e.g., by an interior-point method, for an optimal power $\mathbf{P}[t]$.
6: Set $t := t + 1$;
7: **until P** converge;

Algorithm 5 Joint user clustering and resource allocation in NOMA (JUCRAN)

1: **Step 1: Initialization**
2: The users CSI information is computed by MBS.
3: User classification and computation NOMA clusters are performed utilizing Algorithm 1.
4: Each user is allocated equal transmission power by MBS.
5: Equal resource block allocation to every cluster is performed by MBS.
6: **Step 2: Joint User Clustering and Resource Allocation**
7: **repeat**
8: Perform user clustering update $\boldsymbol{\beta}$ utilizing Algorithm 2.
9: Perform RB allocation update $\boldsymbol{\delta}$ utilizing Algorithm 3.
10: Perform power assignment update $\mathbf{P}$ utilizing Algorithm 4.
11: **until** convergence;

$$\begin{aligned}
&\mathbf{PA2}: \\
&\min_{\mathbf{P}} \prod_{k\in\mathcal{S}} \prod_{j_n\in\mathcal{S}_k} \left(\frac{\sum_{i_m\in\bar{\mathcal{S}}_k(j_n)} P_{i_m}|h_{j_n}|^2 + \tilde{z}_{j_n}}{\sum_{i_m\in\bar{\mathcal{S}}_k(j_n)} P_{i_m}|h_{j_n}|^2 + \tilde{z}_{j_n} + P_{j_n}|h_{j_n}|^2} \right)^{\tilde{\beta}^k_{j_n}\omega_{j_n}} \\
&\text{s.t.: } C_1: \sum_{j_n\in\mathcal{U}} P_{j_n} \le P_T, \\
&C_2: \prod_{j_n\in\mathcal{U}_n} \prod_{k\in\mathcal{S}} \left(\frac{\sum_{i_m\in\bar{\mathcal{S}}_k(j_n)} P_{i_m}|h_{j_n}|^2 + \tilde{z}_{j_n}}{\sum_{i_m\in\bar{\mathcal{S}}_k(j_n)} P_{i_m}|h_{j_n}|^2 + \tilde{z}_{j_n} + P_{j_n}|h_{j_n}|^2} \right)^{\tilde{\beta}^k_{j_n}} \\
&\le 2^{-R_n^{\min}}, \ \forall n.
\end{aligned} \tag{5.22}$$

Define $u^k_{j_n}(\mathbf{P}) = \sum_{i_m\in\bar{\mathcal{S}}_k(j_n)} P_{i_m}|h_{j_n}|^2 + \tilde{z}_{j_n} + P_{j_n}|h_{j_n}|^2$, then according to arithmetic–geometric mean inequality:

$$\begin{aligned}
&u^k_{j_n}(\mathbf{P}) \ge \underline{u}^k_{j_n}(\mathbf{P}) \\
&= \prod_{i_m\in\bar{\mathcal{S}}_k(j_n)} \left(\frac{P_{i_m}|h_{j_n}|^2}{\kappa_{i_m}}\right)^{\kappa_{i_m}} \left(\frac{\tilde{z}_{j_n}}{\lambda_{j_n}}\right)^{\lambda_{j_n}} \left(\frac{P_{j_n}|h_{j_n}|^2}{\gamma_{j_n}}\right)^{\gamma_{j_n}},
\end{aligned} \tag{5.23}$$

where for all $j_n \in \mathcal{S}_k$, $i_m \in \bar{\mathcal{S}}_k(j_n)$, $\kappa_{i_m} = P_{i_m}|h_{j_n}|^2/u^k_{j_n}(\mathbf{P})$, $\lambda_{j_n} = \tilde{z}_{j_n}/u^k_{j_n}(\mathbf{P})$, and $\gamma_{j_n} = P_{j_n}|h_{j_n}|^2/u^k_{j_n}(\mathbf{P})$.

The approximate problem considered from a geometric programs class is given below:

$$\mathbf{PA3}: \min_{\mathbf{P}} \prod_{k\in\mathcal{S}} \prod_{j_n\in\mathcal{S}_k} \left(\frac{\sum\limits_{i_m\in\bar{\mathcal{S}}_k(j_n)} P_{i_m}|h_{j_n}|^2 + \tilde{z}_{j_n}}{\underline{u}^k_{j_n}(\mathbf{P})} \right)^{\tilde{\beta}^k_{jn}\omega_{jn}}$$

$$\text{s.t.: } C_1: \sum_{j_n\in\mathcal{U}} P_{j_n} \le P_T, \tag{5.24}$$

$$C_2: \prod_{j_n\in\mathcal{U}_n} \prod_{k\in\mathcal{S}} \left(\frac{\sum\limits_{i_m\in\bar{\mathcal{S}}_k(j_n)} P_{i_m}|h_{j_n}|^2 + \tilde{z}_{j_n}}{\underline{u}^k_{j_n}(\mathbf{P})} \right)^{\tilde{\beta}^k_{jn}} \le 2^{-R_n^{\min}}, \ \forall n.$$

The transformation of the program of (5.24) into logarithmic variables change can be easily performed [2]. The currently available optimal solution $\mathbf{P}(t)$ will then be utilized in (5.23) for updating the parameters in the next iteration.

Centralized SCA-Based Power Assignment with AGM Approximation

In connection with the above discussion, a power assignment algorithm, Algorithm 4, is proposed. Moreover, the authors have also provided the convergence analysis stating that the solution converges to local optimal solution satisfying *Karush–Kuhn–Tucker (KKT)* conditions of the original problem (5.18).

5.3.4 *Joint User Clustering and Resource Allocation in NOMA: JUCRAN*

This section presents the joint user clustering and resource allocation in NOMA (JUCRAN) algorithm, as illustrated in Algorithm 5. In the initialization phase, the information of classifier users and CSI information of all users is collected by MBS. Then the required number of NOMA clusters is computed using Algorithm 1. This step also includes the initialization of equal number of resource blocks for every NOMA cluster and equal power transmission for all users. In the joint phase 2, all of the three algorithms such as power assignment, RB allocation, and user clustering are iteratively performed till the solution (5.5). Finally, the convergence to a sub-optimal solution of the JUCRAN algorithm for the problem in (5.5) is proved by the authors.

5.4 Simulation Results

5.4.1 Simulation Setting

This section presents the performance evaluation of the joint algorithm JUCRAN for downlink NOMA WNV. For simulation, one BS positioned at the center of the cell. A uniform distribution of the users is considered in a circular cell of range 500 m. The number of resource blocks, each resource block bandwidth, system bandwidth, and maximum transmission power of MBS are considered as $\Omega = 100$, $B = 180$ KHz, 20 MHz, and 40 W, respectively. The maximum transmission power is varied in some simulations for investigation of the trade-off between energy efficiency and throughput. In the simulation, three MVNOs are considered with each having uniformly distributed users along with similar QoS constraints. For all resource blocks, the power of noise is taken equal to 10^{-13} W. The small-scale fading channel coefficients are generated as i.i.d. Rayleigh random variables having unit variance.

Performance comparison of the proposed scheme with OFDMA has been performed. Moreover, the comparison of the proposed scheme JUCRAN for three different users cases (such as weak users, normal users, and strong users) denoted by JUCRAN-3 and given in Algorithm 1 is also performed with a scheme JUCRAN-2 that uses two users classes. The usage of two users cases is similar to a proposal in [1]. All the results are obtained by taking an average of the 100 simulations that are performed for random topologies.

5.4.2 Numerical Results

The number of iterations needed by the JUCRAN-3 and JUCRAN-2 schemes versus network users number is shown in Fig. 5.3. It is clear from Fig. 5.3 that the number of iterations for JUCRAN-2 is higher than JUCRAN-3. Further, the number of iterations has an increased trend for the rise in network users. In JUCRAN-3, the iterations needed for convergence are less due to the fact that quota of each cluster is taken 3 which tended to accept more users. Furthermore, the number of clusters that are created in JUCRAN-2 is more than for JUCRAN-3 scheme which causes the accept-reject process to use more iterations because of the high number of clusters. The average number of cluster formations versus network users is plotted in Fig. 5.4, which reveals that the number of clusters for JUCRAN-2 is always higher than JUCRAN-3. This trend is due to the fact that a number of users per class in JUCRAN-3 are lower than JUCRAN-2.

The convergence of the proposed joint scheme is shown in Fig. 5.5. The performance is evaluated for three different instances such as 50, 30, and 10 users. It is clear from Fig. 5.5 that the proposed JUCRAN converges quickly for limited iterations. Moreover, the relative utility of JUCRAN-3 is higher than JUCRAN-2.

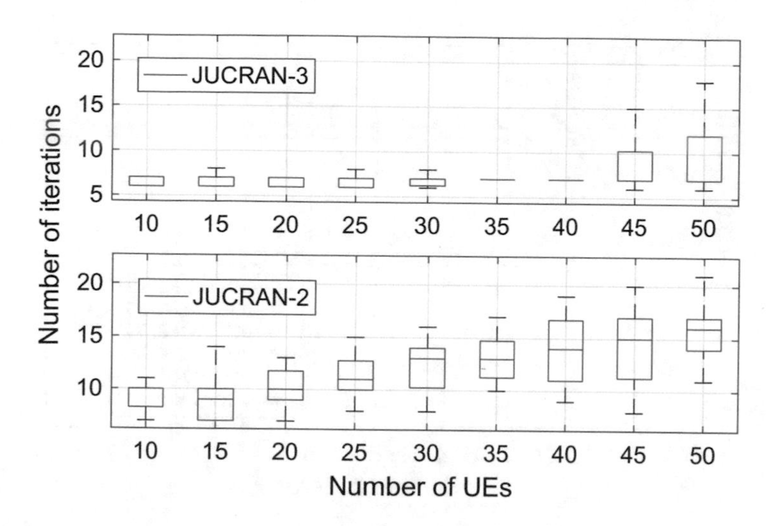

Fig. 5.3 The number of iteration of matching algorithm with two schemes: JUCRAN-2 and JUCRAN-3

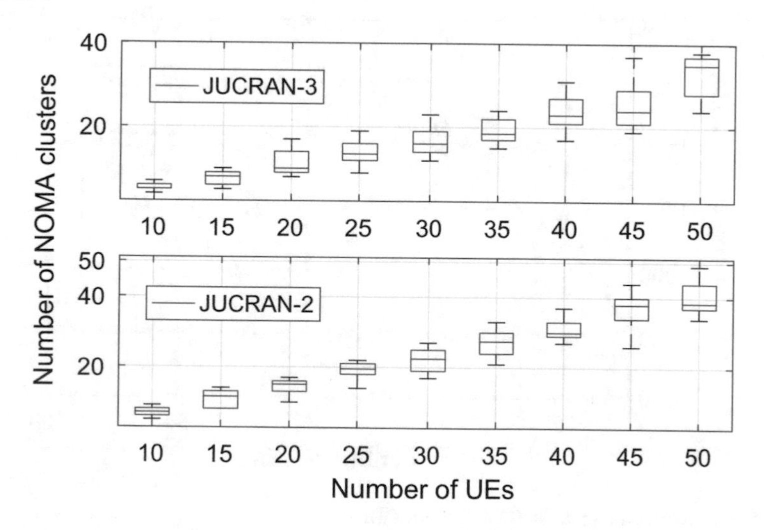

Fig. 5.4 The number of NOMA cluster with two schemes: JUCRAN-2 and JUCRAN-3

This is due to the fact that JUCRAN-3 has the ability to accommodate more users than JUCRAN-2, which enable JUCRAN-3 to fulfill the minimum rate for more users. In Fig. 5.6, the network throughput is plotted versus the network's users. The total throughput is evaluated through the average total sum-rate over the different number of users. The throughput of both JUCRAN-3 and JUCRAN-3 increases with an increase in the number of users until the number of users gets saturated. For the number of users more than 50, it is observed that the total throughput shows a trend of continuous increase because of the multi-diversity gain. However, this

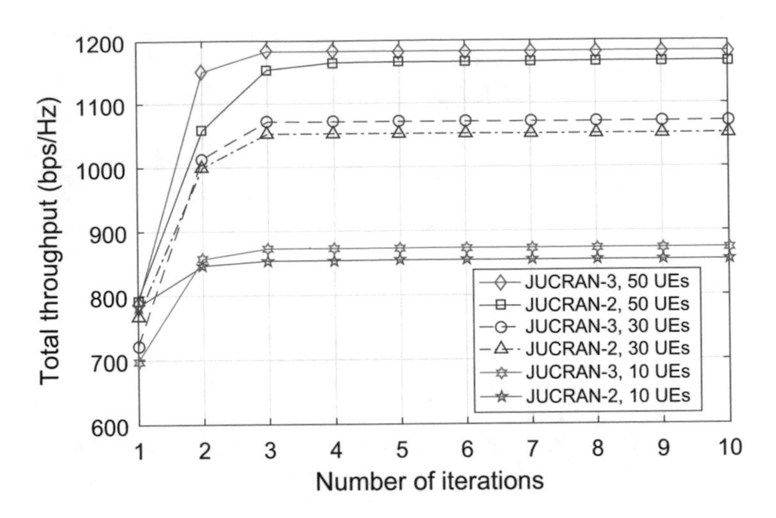

Fig. 5.5 JUCRAN-3 and JUCRAN-2 convergence

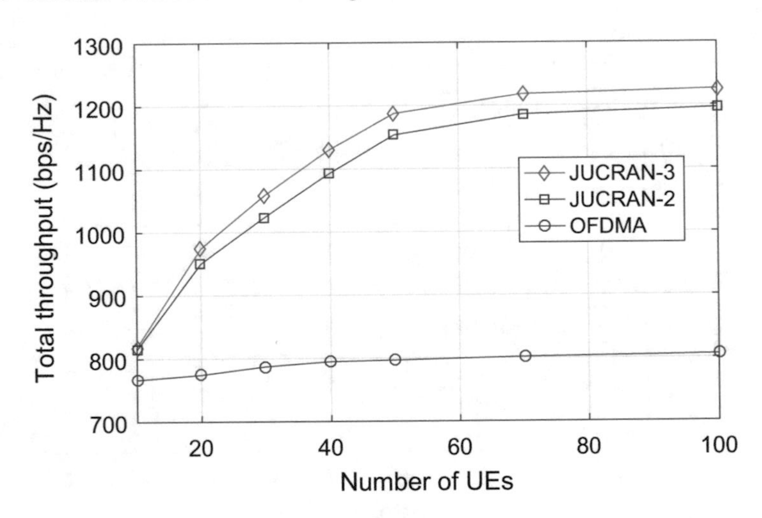

Fig. 5.6 Network throughput JUCRAN vs OFDMA

increase in throughput becomes slow and eventually saturated when the number of users gets significantly large. The performance of JUCRAN-2 and JUCRAN-3 is almost comparable, but better than OFDMA significantly. From Fig. 5.6, it is clear that InP cannot obtain a significant advantage of resource allocation for the case OFDMA.

In Fig. 5.7, the network throughput versus maximum transmit power for 50 users is plotted. A throughput of both JUCRAN-2 and JUCRAN-3 is more than OFDMA

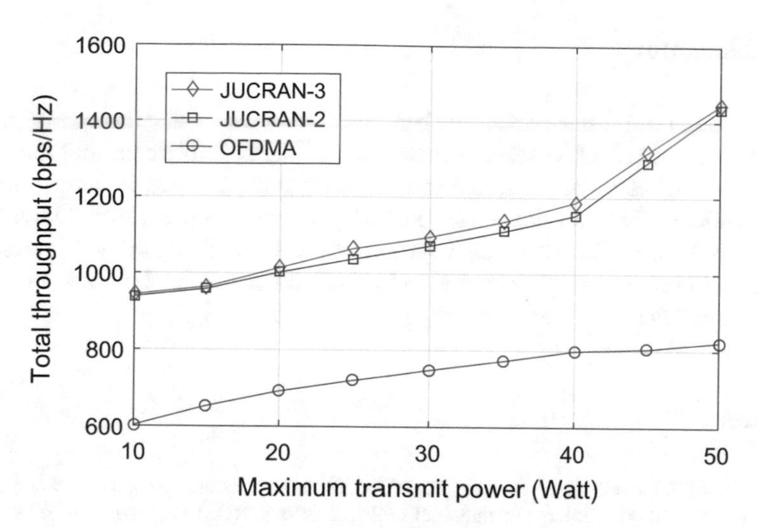

Fig. 5.7 Network throughput vs maximum transmit power with 50 users

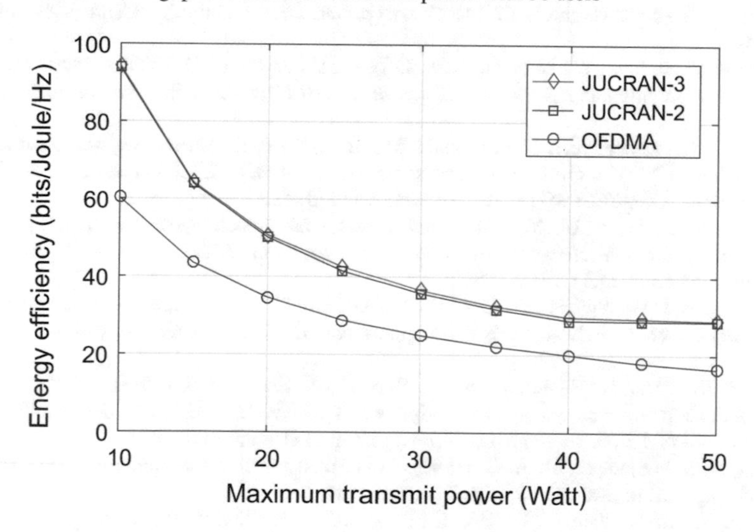

Fig. 5.8 Energy efficiency vs maximum transmit power with 50 users

for all values of transmit power. Other than that, the throughput of both JUCRAN-3 and JUCRAN-2 increases with an increase in MBS transmission power. Finally, Fig. 5.8 shows the energy efficiency versus maximum transmit power for 50 users. It is clear from Fig. 5.8 that energy efficiency decreases with an increase in MBS transmit power budget. Both JUCRAN-2 and JUCRAN-3 outperformed OFDMA.

5.5 Summary

In this chapter, we discuss about the benefits achieved by using the non-orthogonal multiple access (NOMA) scheme over the traditional orthogonal multiple access scheme in a wireless virtualization setting. User clustering and resource allocation are considered the vital challenges in NOMA that requires to be addressed by efficient schemes. Furthermore, we surveyed a work that jointly addresses the problem of user clustering, resource, and power allocation in wireless virtualization setting to enhance the overall network performance.

References

1. Ali, M. S., Tabassum, H., & Hossain, E. (2016). Dynamic user clustering and power allocation for uplink and downlink non-orthogonal multiple access (NOMA) systems. *IEEE Access, 4*, 6325–6343.
2. Boyd, S., & Vandenberghe, L. (2004). *Convex optimization*. Cambridge: Cambridge University Press.
3. Chiang, M., Tan, C. W., Palomar, D. P., O'Neill, D., & Julian, D. (2005). Power control by geometric programming. *Resource Allocation in Next Generation Wireless Networks, 5*, 289–313.
4. Dawadi, R., Parsaeefard, S., Derakhshani, M., & Le-Ngoc, T. (2016). Power-efficient resource allocation in NOMA virtualized wireless networks. In *2016 IEEE Global Communications Conference (GLOBECOM)* (pp. 1–6). Piscataway: IEEE.
5. Di, B., Song, L., & Li, Y. (2016). Sub-channel assignment, power allocation, and user scheduling for non-orthogonal multiple access networks. *IEEE Transactions on Wireless Communications, 15*(11), 7686–7698.
6. Gu, Y., Saad, W., Bennis, M., Debbah, M., & Han, Z. (2015). Matching theory for future wireless networks: Fundamentals and applications. *IEEE Communications Magazine, 53*(5), 52–59.
7. Ho, T. M., Tran, N. H., Kazmi, S. A., Han, Z., & Hong, C. S. (2018). Wireless network virtualization with non-orthogonal multiple access. In *NOMS 2018-2018 IEEE/IFIP Network Operations and Management Symposium* (pp. 1–9). Piscataway: IEEE.
8. Roth, A. & Sotomayor, M. A. O. (1992). Two-Sided Matching: A Study in Game-Theoretic Modeling and Analysis. Cambridge University Press.
9. Kazmi, S. M. A., Tran, N. H., Ho, T. M., Manzoor, A., Niyato, D., & Hong, C. S. (2018). Coordinated device-to-device communication with non-orthogonal multiple access in future wireless cellular networks. *IEEE Access, 6*, 39860–39875.
10. Kazmi, S. A., Tran, N. H., Saad, W., Han, Z., Ho, T. M., Oo, T. Z., et al. (2017). Mode selection and resource allocation in device-to-device communications: A matching game approach. *IEEE Transactions on Mobile Computing*(11), 3126–3141.
11. Liang, C., & Yu, F. R. (2015). Wireless network virtualization: A survey, some research issues and challenges. *IEEE Communications Surveys & Tutorials, 17*(1), 358–380.
12. Liang, C., & Yu, F. R. (2015). Wireless virtualization for next generation mobile cellular networks. *IEEE Wireless Communications, 22*(1), 61–69.
13. Marks, B. R., & Wright, G. P. (1978). A general inner approximation algorithm for nonconvex mathematical programs. *Operations Research, 26*(4), 681–683.
14. Saito, Y., Kishiyama, Y., Benjebbour, A., Nakamura, T., Li, A., & Higuchi, K. (2013). Non-orthogonal multiple access (NOMA) for cellular future radio access. In *2013 IEEE 77th Vehicular Technology Conference (VTC Spring)* (pp. 1–5). Piscataway: IEEE.

15. Song, L., Li, Y., Ding, Z., & Poor, H. V. (2017). Resource management in non-orthogonal multiple access networks for 5G and beyond. *IEEE Network, 31*(4), 8–14.
16. Wen, H., Tiwary, P. K., & Le-Ngoc, T. (2013). Current trends and perspectives in wireless virtualization. In *2013 International Conference on Selected Topics in Mobile and Wireless Networking (MoWNeT)* (pp. 62–67). New York: IEEE.
17. Zhu, K., & Hossain, E. (2016). Virtualization of 5G cellular networks as a hierarchical combinatorial auction. *IEEE Transactions on Mobile Computing, 15*(10), 2640–2654.

Chapter 6
Network Slicing: Cache and Backhaul Resource Allocation

6.1 Introduction

Traditionally in cellular networks radio resources were only considered as a performance bottleneck. Therefore, a number of solutions were devised to only cater the radio resource allocation challenge. A similar trend has been followed for deriving solutions for network slicing as well in which only radio resources were sliced to fulfill the end user requirements which has been discussed in the previous chapters. However, the proliferation of end users and novel applications have also imposed limitations on other network resources such as backhaul and cache spaces. In this chapter, we would discuss two novel solutions which consider these resources along with the radio resources to build networking slices.

6.2 Network Slicing with Backhaul Constraints

In the first work [6], the backhaul resources are considered for slicing the network. This work mainly dealt with the network virtualization targeting the uplink of a cellular network. They developed a novel technique enchaining the spectrum and infrastructure efficiency. They designed the system capable of maximizing the MVNO profit along with guaranteeing the users QoS requirements and InP's backhaul constraints. They used slice isolation-based approach for uplink of a cellular network, for finding the sub-optimal decision on slice and transmit power allocation. Then, they designed a distributed algorithm based on Lagrangian relaxation, and for the convergence of sub-optimal point they utilized matching game technique.

S. M. A. Kazmi et al., *Network Slicing for 5G and Beyond Networks*,
https://doi.org/10.1007/978-3-030-16170-5_6

6.2.1 System Model

Consider the uplink transmission of the OFDMA-based virtualized cellular network as illustrated in Fig. 6.1. The set $\mathcal{B} = \{1, 2, \ldots, B\}$ of infrastructure providers (InPs) provides the network resources to mobile virtual network operators (MVNO) on rent. The system model consists of a small area covered by overlapping small cells of multiple InPs. The set $\mathcal{I} = \{1, 2, \ldots, I\}$ of service providers (SP) are provided with the virtual resources by the MVNO. Every service provider $i \in \mathcal{I}$ has its own subscribers set $\mathcal{U}^i = \{1, 2, \ldots, U_i\}$. The users set of all service providers is represented by $\mathcal{U}$. The infrastructure provider b has subcarriers set $\mathcal{L}_b = \{1, 2, \ldots, L_b\}$, that is further transformed into $\mathcal{C}_b = \{1, 2, \ldots, C_b\}$ chunks. Every chunk is formed by aggregating $\mathcal{L}_{b,c} = \{1, 2, \ldots, L_b/C_b\}$ subcarriers. The bandwidth of narrowband orthogonal subcarriers is W. Apart from that, no interference is assumed between the small cells of different and same infrastructure providers.

The set that consists of $\mathcal{S}$ slices is used for isolation of the infrastructure services. A pair of chunk and base station is utilized for unique determination of a slice. The slice allocation matrix is denoted by $\boldsymbol{\alpha} = [\alpha^u_{b,c}]_{|\mathcal{U}|\times(|\mathcal{C}_b|\times B)}$. The binary variable

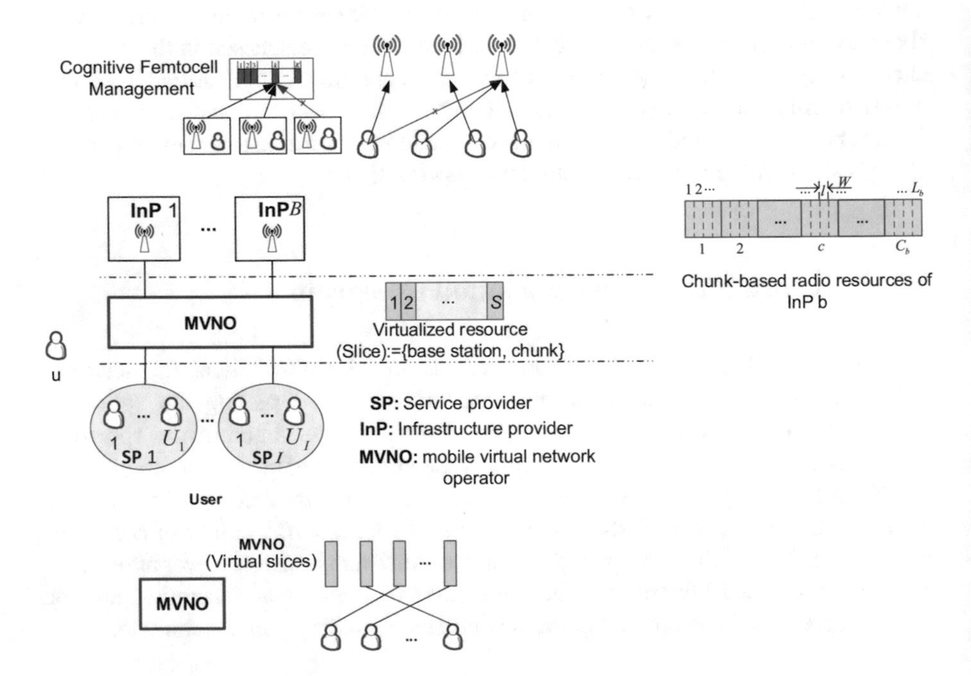

Fig. 6.1 Virtualized cellular network hierarchical model

$\alpha_{b,c}^{u}$ is used to indicate whether the slice $\{b, c\}$ ($b \in \mathcal{B}, c \in \mathcal{C}_b$) is allocated to user u or not (such as $\alpha_{b,c}^{u} = 1$ if the slice is allocated and $\alpha_{b,c}^{u} = 0$ otherwise).

Problem Formulation

The data rate of a user u associated with service provider i that performs transmission scheduled by MVNO is given by:

$$R_u^i(\boldsymbol{\alpha}, \boldsymbol{P}) = \sum_{b \in \mathcal{B}} \sum_{c \in \mathcal{C}_b} \alpha_{b,c}^{u} r_{b,c}^{u}(\boldsymbol{P}_{b,c}^{u}), \tag{6.1}$$

where $r_{b,c}^{u}(\boldsymbol{P}_{b,c}^{u}) = \sum_{l \in \mathcal{L}_{b,c}} W \log_2(1+\gamma_{b,c}^{u,l} P_{b,c}^{u,l})$ denote the data rate (chunk based) of user u that has association with chunk c and small-cell base station b; $\gamma_{b,c}^{u,l} = \frac{g_{b,c}^{u,l}}{\sigma_b^2}$; W denote the subcarrier $l \in \mathcal{L}_b, \forall b$ bandwidth; the instantaneous channel gain is denoted by $g_{b,c}^{u,l}$; the transmit power vector for chunk c subcarriers is denoted by $\boldsymbol{P}_{b,c}^{u} = [P_{b,c}^{u,l}]_{1 \times |\mathcal{L}_{b,c}|}; \boldsymbol{P} = [\boldsymbol{P}_{b,c}^{u}]_{|\mathcal{U}| \times (|\mathcal{C}_b| \times B)}$ represents the transmit power matrix.

Consider the user u that is subscribed to the service provider i, then the constraint for guaranteeing the required minimum rate $R_{u,\min}^{i}$ is given by:

$$R_u^i(\boldsymbol{\alpha}, \boldsymbol{P}) \geq R_{u,\min}^{i}, \ \forall u \in \mathcal{U}, \forall i \in \mathcal{I}. \tag{6.2}$$

The aggregated data rate of users must satisfy the constraint given below to avoid congestion at the backhaul links of infrastructure provider small-cell base station:

$$\sum_{i \in \mathcal{I}} \sum_{u \in \mathcal{U}_i : \alpha_{b,c}^{u} = 1} R_u^i(\boldsymbol{\alpha}, \boldsymbol{P}) \leq Z_{b,\text{bh}}, \ \forall b \in \mathcal{B}, \ \forall c \in \mathcal{C}_b, \tag{6.3}$$

where the small-cell base station b backhaul capacity is $Z_{b,\text{bh}} \geq 0$. The MVNO network utility attained by allocation of the transmit power and slice allocation to service providers users is given by:

$$U_{\text{MVNO}}(\boldsymbol{\alpha}, \boldsymbol{P}) = U^{\text{rev}}(\boldsymbol{\alpha}, \boldsymbol{P}) - U^{\text{cost}}(\boldsymbol{\alpha}, \boldsymbol{P}), \tag{6.4}$$

where $\varphi_{b,c}^{\text{slice}}$ denote the slice unit cost assigned by infrastructure provider b for small-cell base station b chunk c; φ_b^{bh} backhaul unit price assigned by infrastructure provider b regarding small-cell base station b; φ_i^{sp} represent the payment taken by MVNO from the service provider i ; $U^{\text{cost}}(\boldsymbol{\alpha}, \boldsymbol{P}) = \sum_{b \in \mathcal{B}} \sum_{c \in \mathcal{C}_b} (\varphi_b^{\text{bh}} \alpha_{b,c}^{u} r_{b,c}^{u}(\boldsymbol{P}_{b,c}^{u}) + \varphi_{b,c}^{\text{slice}} \alpha_{b,c}^{u})$ denote the total cost occurred when the infrastructure provider gives its resources on lease to MVNO; $U^{\text{rev}}(\boldsymbol{\alpha}, \boldsymbol{P}) = \sum_{i \in \mathcal{I}} \sum_{u \in \mathcal{U}^i} \varphi_i^{\text{sp}} R_u^i(\boldsymbol{\alpha}, \boldsymbol{P})$ denote the revenue of MVNO resulted from giving the resources to service providers. The problem is formulated mathematically as follows:

(OP):

$$\max_{(\boldsymbol{\alpha}, \boldsymbol{P})} \quad U_{\text{MVNO}}(\boldsymbol{\alpha}, \boldsymbol{P}) \tag{6.5}$$

$$\text{s.t. } (6.2), (6.3),$$

$$\sum_{b \in \mathcal{B}} \sum_{c \in \mathcal{C}_b} \alpha^u_{b,c} \sum_{l \in \mathcal{L}_c} P^{u,l}_{b,c} \leq \bar{P}_u, \ \forall u \in \mathcal{U}, \tag{6.6}$$

$$P^{u,l}_{b,c} \geq 0, \ \forall b \in \mathcal{B}, \forall c \in \mathcal{C}_b, \forall u \in \mathcal{U}, \tag{6.7}$$

$$\alpha^u_{b,c} \in \Pi_{\boldsymbol{\alpha}}, \ \forall b \in \mathcal{B}, \forall c \in \mathcal{C}_b, \forall u \in \mathcal{U}, \tag{6.8}$$

where (6.6) puts constraints on the total transmit power ($\bar{P}_u$); $\Pi_{\boldsymbol{\alpha}}$ is the following non-convex set:

$$\sum\nolimits_{u \in \mathcal{U}} \alpha^u_{b,c} \leq 1, \ \forall c \in \mathcal{C}_b, \ \forall b \in \mathcal{B}, \tag{6.9}$$

$$\sum\nolimits_{b \in \mathcal{B}} \sum\nolimits_{c \in \mathcal{C}_b} \alpha^u_{b,c} \leq 1, \ \forall u \in \mathcal{U}, \tag{6.10}$$

$$\sum\nolimits_{u \in \mathcal{U}} \sum\nolimits_{c \in \mathcal{C}_b} \alpha^u_{b,c} \leq 1, \ \forall b \in \mathcal{B}, \tag{6.11}$$

$$\sum\nolimits_{u \in \mathcal{U}} \sum\nolimits_{b \in \mathcal{B}} \alpha^u_{b,c} \leq 1, \ \forall c \in \mathcal{C}_b, \tag{6.12}$$

$$\alpha^u_{b,c} = \{0, 1\}, \forall u \in \mathcal{U}, \ \forall b \in \mathcal{B}, \ \forall c \in \mathcal{C}_b. \tag{6.13}$$

Constraint (6.9) restricts the allocation of a slice to at most user. Constraint (10) restricts the assignment of user to at most one slice. Equations (6.11) and (6.12) denote the isolation of the slices whose unique criteria is set by MVNO. The nature of the problem (**OP**) is mixed-integer non-convex optimization problem, whose computational solution is intractable. A solution (sub-optimal) is then proposed to solve the problem (**OP**) after Lagrangian relaxation (Table 6.1).

6.2.2 *Solution Approach: Joint Slice and Power Allocation*

The problem (**OP**) partial Lagrangian is attained using augmentation between the weighted sum of constraints (6.2), (6.3), (6.6) and objective function as follows:

$$L(\boldsymbol{\alpha}, \boldsymbol{P}, \boldsymbol{\lambda}, \boldsymbol{\beta}) = U_{\text{MVNO}}(\boldsymbol{\alpha}, \boldsymbol{P}) + \sum_{i \in \mathcal{I}} \sum_{u \in \mathcal{U}^i} \lambda_u (R^i_u(\boldsymbol{\alpha}, \boldsymbol{P}) - R^i_{u,\min})$$

Table 6.1 Some notations used in this paper

$\mathcal{B}$	Set of InPs
$\mathcal{I}$	Set of virtual resources given by MVNO to SPs
$\mathcal{U}_m$	Set of customers connected with SPs
$\mathcal{L}_b$	Set of radio resources subcarriers owned by InPs
$\mathcal{C}_b$	Set of subcarriers chunks
α	Slice allocation matrix
$\alpha^{\mu}_{b,c}$	Binary indicator matrix
R^{i}_{μ}	Data rate
σ^2_b	Background noise
W	Bandwidth
$P^{\mu}_{b,c}$	Transmission power vector
$Z_{b,bh}$	Predefined Backhaul capacity
ϕ^{sp}_i	The payment (in units/Mbps) of each SP i to the MVNO
ϕ^{bh}_b	Unit price (in units/Mbps) of the Backhaul set by InP
$L_{\alpha,P,\lambda,\beta}$	Lagrangian function
λ	Lagrangian nonnegative multiplier
β	Lagrangian nonnegative multiplier
μ	Lagrangian nonnegative multiplier
$\omega^{\mu}_{b,c}$	Used in Lagrangian dual function
ϕ^{k}_u	Utility function

$$-\sum_{b\in\mathcal{B}}\beta_b\Big(\sum_{i\in\mathcal{I}}\sum_{u\in\mathcal{U}_i}R^i_u(\boldsymbol{\alpha},\boldsymbol{P})-Z_{b,\mathrm{bh}}\Big)$$

$$-\sum_{i\in\mathcal{I}}\sum_{u\in\mathcal{U}_i}\mu_u\Big(\sum_{b\in\mathcal{B}}\sum_{c\in\mathcal{C}_b}\alpha^u_{b,c}\sum_{l\in\mathcal{L}_c}P^{u,l}_{b,c}-\bar{P}_u\Big). \tag{6.14}$$

where $\boldsymbol{\mu}=[\mu_u]_{1\times(|\mathcal{U}|)}$, $\boldsymbol{\lambda}=[\lambda_u]_{1\times(|\mathcal{U}|)}$, and $\boldsymbol{\beta}=[\beta_b]_{1\times B}$ denote the Lagrangian nonnegative multipliers of the constraints (6.6), (6.2), and (6.3), respectively.

The problem (**OP**) has Lagrangian dual function as:
(**D**)

$$\max_{(\boldsymbol{\alpha},\boldsymbol{P})}\sum_{i\in\mathcal{I}}\sum_{u\in\mathcal{U}^i}\sum_{b\in\mathcal{B}}\sum_{c\in\mathcal{C}_b}\alpha^u_{b,c}\Big[\Omega^u_{b,c}(\boldsymbol{P}^u_{b,c})-\varphi^{\mathrm{slice}}_{b,c}\Big] \tag{6.15}$$

s.t. (6.7), (6.8),

where $\Omega^u_{b,c}(\boldsymbol{P}^u_{b,c}) = (\varphi^{\text{sp}}_i - \varphi^{\text{bh}}_b + \lambda_u - \beta_b) r^u_{b,c}(\boldsymbol{P}^u_{b,c}) - \mu_u \sum_{l \in \mathcal{L}_{b,c}} P^{u,l}_{b,c}$.

The optimal power is determined using KKT condition [1] irrespective of Lagrangian multiplier values and the allocation of slices $\boldsymbol{\alpha}$ through performing the first order derivative of $\Omega^u_{b,c}(\boldsymbol{P}^u_{b,c})$ with respect to $P^{u,l}_{b,c}$ as:

$$P^{u,l*}_{b,c} = \left[\frac{\varphi^{\text{sp}}_i - \varphi^{\text{bh}}_b + \lambda_u - \beta_b}{(\ln 2/W)\mu_u} - \frac{1}{\gamma^{u,l}_{b,c}} \right]^+, \tag{6.16}$$

where $(x)^+ = \max(x, 0)$.

The problem (**D**) can be reduced to a maximum weighted matching given the power allocation in (6.16) as:

(D-1)

$$\max_{(\boldsymbol{\alpha}, \boldsymbol{P})} \sum_{i \in \mathcal{I}} \sum_{u \in \mathcal{U}^i} \sum_{b \in \mathcal{B}} \sum_{c \in \mathcal{C}_b} \alpha^u_{b,c} \left[\Omega^u_{b,c}(\boldsymbol{P}^{u*}_{b,c}) - \varphi^{\text{slice}}_{b,c} \right] \tag{6.17}$$

$$\text{s.t.} \quad (6.9), (6.10), (6.13).$$

The Hungarian algorithm [5] is used to perform optimal slice allocation and $[\Omega^u_{b,c}(\boldsymbol{P}^{u*}_{b,c}) - \varphi^{\text{slice}}_{b,c}]$ is used to weight slice $\{b, c\}$ for user u.

A distributed algorithm (JSPA-HSA algorithm) is presented in Algorithm 1 for solving JSPA problem. The Lagrangian multipliers optimal value is yielded through projected gradient-descent method (given $\boldsymbol{\alpha}$ and $\boldsymbol{P}$) [1] and according to (6.20), (6.19), and (6.18) with positive step sizes $s_3(t)$, $s_2(t)$, and $s_1(t)$. The gradient-based standard technique [1] can be utilized to prove the convergence of the JSPA-HSA algorithm. Apart from that, the MVNO requires global information about $\Omega^u_{b,c}(\boldsymbol{P}^u_{b,c})$ for all slices [7]. To reduce the high associated complexity with this algorithm, a distributed algorithm that has lower complexity is proposed.

Matching-Based Low-Complexity Algorithm

This section presents solution having low complexity to tackle the problem (**OP**). A two-sided matching game $(\mathcal{U}, \mathcal{S}, \succ_{\mathcal{U}}, \succ_{\mathcal{S}})$ is used for slice allocation [3]. Here, $\succ_{\mathcal{S}} = \{\succ_{b,c}\}_{\{b,c\} \in \mathcal{S}}$ and $\succ_{\mathcal{U}} = \{\succ_u\}_{u \in \mathcal{U}}$ represent the preference relations of the slices and users, respectively. The two-sided matching game is defined as a function $\mu \colon \mathcal{U} \mapsto \mathcal{S}$ such that:

(1) $u = \mu(\{b, c\}) \leftrightarrow \{b, c\} = \mu(u), \forall u \in \mathcal{U}, \{b, c\} \in \mathcal{S}$;
(2) $|\mu(\{b, c\})| \leq 1$ and $|\mu(u)| \leq 1, u \in \mathcal{U}, \{b, c\} \in \mathcal{S}$.

The relation $\{b, c\} \succ_u \{b, c\}'$ ($\{b, c\}, \{b, c\}' \in \mathcal{S}$) denotes preference of slice $\{b, c\}$ over slice $\{b, c\}'$ in matching μ. Apart from that, $u \succ_{\{b,c\}} u'$ $(u, u' \in \mathcal{U})$ denote the slice $\{b, c\}$ preference of user u over user u'. The blocking pair in case of existence of $\{b, c\} \succ_u \{b, c\}'$ or $u \succ_{\{b,c\}} u'$, $\forall u, b, c, i$ is given by $(u, \{b, c\})$. The

Algorithm 1 JSPA-HSA: JSPA with Hungarian-based slice allocation

1: Initialization: $\mathcal{I}$, $\mathcal{B}$, $\mathcal{C}_b$, $\mathcal{U}_i$, $\boldsymbol{P}^{(0)}$, $\boldsymbol{\lambda}^{(0)}$, $\boldsymbol{\mu}^{(0)}$, and $\boldsymbol{\beta}^{(0)}$.
2: Repeat:
3: **Power allocation phase**:
4: *At the subscribed user u:
5: Update λ_u as:

$$\lambda_u(t+1) = [\lambda_u(t) - s_1(t)(R_u^i(\boldsymbol{\alpha}, \boldsymbol{P}) - R_u^{\min})]^+; \tag{6.18}$$

6: Update μ_u as:

$$\mu_u(t+1) = \left[\mu_u(t) - s_2(t)\left(\sum_{b\in\mathcal{B}}\sum_{c\in\mathcal{C}_b}\alpha_{b,c}^u\sum_{l\in\mathcal{L}_c}P_{b,c}^{u,l} - \bar{P}_u\right)\right]^+; \tag{6.19}$$

7: Update transmit power $P_{b,c}^{u,l}(t+1)$ by (6.16);
8: *At the SBS b:
9: Update congested backhaul link price $\beta_b(t+1)$:

$$\beta_b(t+1) = \left[\beta_b(t) + s_3(t)\left(\sum\nolimits_{i\in\mathcal{I}}\sum\nolimits_{u\in\mathcal{U}_i} R_u^i(\boldsymbol{\alpha}, \boldsymbol{P}) - Z_{b,\text{bh}}\right)\right]^+; \tag{6.20}$$

10: **Slice allocation phase**:
11: *At the MVNO:
12: Update $\alpha_{b,c}^u(t+1)$ using the Hungarian algorithm to maximize (6.17).
13: Until $|\lambda_u(t+1) - \lambda_u(t)| \le \epsilon_1$, $|\mu_u(t+1) - \mu_u(t)| \le \epsilon_2$, and $|\beta_b(t+1) - \beta_b(t)| \le \epsilon_3$ are simultaneously satisfied.

---

utility functions $\phi_{\{b,c\}}(u)$ and $\phi_u(\{b,c\})$ form the preference relations $\succ_{\{b,c\}}$ and $\succ_u$ for MVNO and the users, respectively.

The utility value $\phi_u(\{b,c\}) = \Omega_{b,c}^u(\boldsymbol{P}_{b,c}^u)$ is used for estimation of users u utilities for slices in the proposed two-sided matching game. Each user wants to maximize the value of the utility through bidding of the slice $\{b,c\}^* := \arg\max_{\{b,c\}\in\mathcal{S}} \Omega_{b,c}^u(\boldsymbol{P}_{b,c}^u)$ in its list of preferences. The MVNO wants to perform maximization of its utility in reaction to users requests of getting the slices as follows:

$$\phi_u(k) = \Omega_{b,c}^u(\boldsymbol{P}_{b,c}^u) - \varphi_{b,c}^{\text{slice}}. \tag{6.21}$$

The strategy for distributed slice allocation is given in Algorithm 1 (MSA Algorithm) to maximize the objective function (6.17). The upper bound $\mathcal{O}(|\mathcal{U}|^2(|\mathcal{S}|-1))$ can be utilized to determine the complexity of Algorithm 1. The value $\phi_{\{b,c\}}(u)$ is captured by the MSA algorithm in the processes of rejection and acceptance on a slice. This type of execution has an increasing effect on objective value of (6.17). Therefore, the problem **(D-1)** maximal value is attained through convergence of

Algorithm 2 MSA: matching-based slice allocation

1: **while** $\sum_{\forall u,\{b,c\}} b_{u\rightarrow\{b,c\}} \neq 0$ or convergence not achieved **do**
2: *At the subscribed users*:
3: Send a bid for the slice $\{b, c\}^* = \arg\max_{\{b,c\}\in\succ_u} \phi_u(\{b, c\})$.
4: *At the MVNO*:
5: Construct $\succ_{\{b,c\}}$ based on (6.21).
6: Update $\{b, c\}^* = \mu(\{b, c\})|u^* = \arg\max_{u\in\succ_{\{b,c\}}} \phi_{\{b,c\}}(u)\}$.
7: Update the rejected user lists on the slices and the preference $\succ_u$.
8: **end while**

MSA algorithm. However, a sub-optimal solution is attained due to the fact that execution of the MSA algorithm is interrupted at stable matching μ^*. Using the above analysis, a low-complexity algorithm JSPA-MSA that has distributed nature is proposed for solution of the problem (**OP**).

6.2.3 Simulation Results

To evaluate the performance of the proposed distributed algorithm, $B = 3$ infrastructure providers are considered. Each infrastructure provider small-cell base station has signal bandwidth of 3MHz and coverage of small-cell base station of 100 m. Every small-cell base station consists of 10 chunks with each containing 12 subcarriers. Every subcarrier has bandwidth $W = 15$ kHz. The two service providers (each has 10 users) are served by the MVNO by taking the resources from the infrastructure provider on rent. The minimum target rate of $200 \times i$ kbps ($i = 1, 2$) is used for the service provider i. The independent and identical Rayleigh distributed random variables are used to model the small-scale channel gains. The $L_d = 38.46 + 20\log_{10}(d)$ represent the large-scale path loss in dB. The power of noise is taken equal to -174 dBm/Hz. The power $P_u^{\max} = 100$ mW is set for user u. For service provider 2 and service provider 1, we use $\varphi_2^{\mathrm{sp}} = 3.5$ and $\varphi_1^{\mathrm{sp}} = 2.5$ units/Mbps, respectively. The prices of backhaul for infrastructure providers 3, 2, and 1 are set to 0.6, 0.4, and 0.2 units/slice, respectively. Along with this, error tolerance $\epsilon = 10^{-3}$ is taken for all algorithms (Table 6.2).

Figures 6.2 and 6.3 illustrate the fast convergence (such as in few iterations) of the proposed algorithms. The gap between the JSPA-HSA scheme and the JSPA-MSA scheme is 3.98% as indicated by the results. Figure 6.4 illustrates the comparison of the profits achieved using the proposed algorithms by MVNO. A baseline Max-Rate algorithm is also considered in Fig. 6.4. Figure 6.5 illustrates the network utility versus service providers number for 10 Mbps backhaul rate. It is clear from Fig. 6.5 that the network utilities get improved with an increase in number of users. The proposed algorithms outperformed Max-Rate.

Table 6.2 Default simulation parameters

Simulation parameters	Values
Number of infrastructure providers	$B = 3$
Number of SBS in each InPs	1
Signal bandwidth	3 MHz
Chunks in each SBS	10
Subcarriers in each chunk	12
Bandwidth of subcarrier W	15 KHz
MVNO served to SPs by rental resources of InPs	2
Number of customers in each SP	10
Minimum target rate set by SP	200*i kbps
Channel gain	Rayleigh random variables with unit mean
Path loss in distance d	38:46 + 20 log10(d) dB
Noise power	−174 dBm/Hz
Maximum power for each customer u ($P_u^{\max}$)	100 mW
Prices for SPs set by MVNO ϕ_1^{sp} and ϕ_2^{sp}	2.5 and 3.5 units/Mbps respectively
Prices for backhaul InPs 1, 2 and 3	0.2, 0.4 and 0.6 units/Mbps respectively
Prices selected for slices of InPs 1, 2, and 3	0.1, 0.2, and 0.3 units/slice respectively
Error tolerance ϵ	10^{-3}

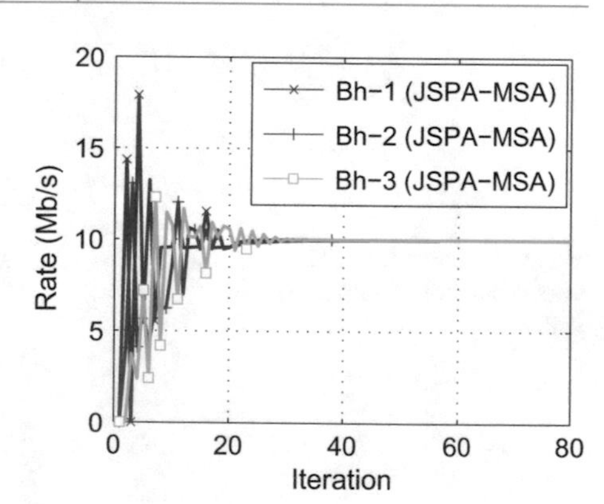

Fig. 6.2 Evaluation results of Rate with $Z_{b,\mathrm{bh}} = 10$ Mbps

6.3 Network Slicing with Backhaul and Cache Allocation

Next, we would discuss the other proposal [10] in which network slicing is done by considering both the backhaul and caching resources.

6.3.1 System Model

Consider Fig. 6.6, which shows architecture that consists of tenants set $\mathcal{N}$ that share network physical resources such as cache enabled base station along with

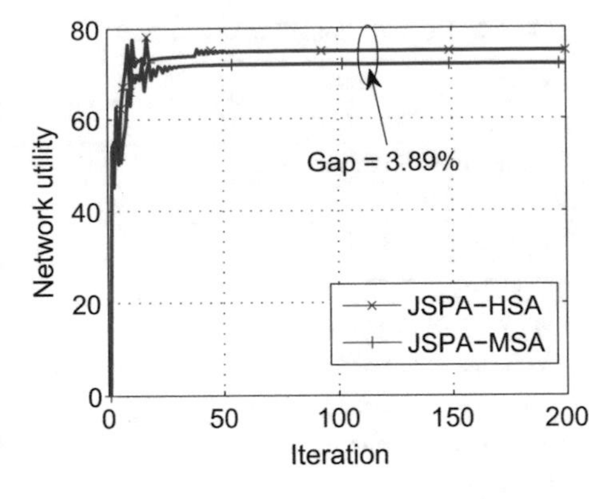

Fig. 6.3 Evaluation results of Network utility with $Z_{b,\text{bh}} = 10\,\text{Mbps}$

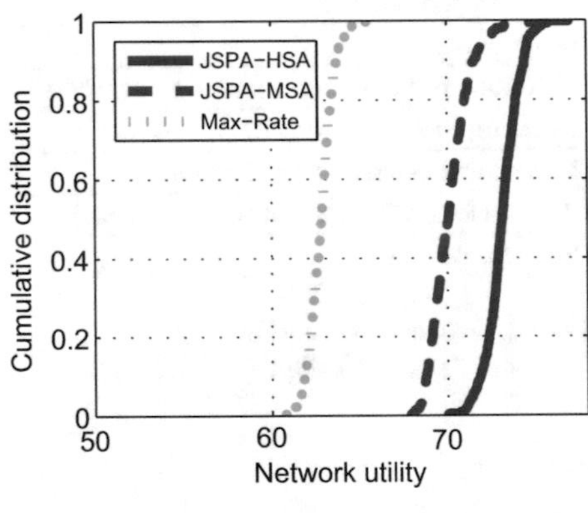

Fig. 6.4 CDF of network utilities

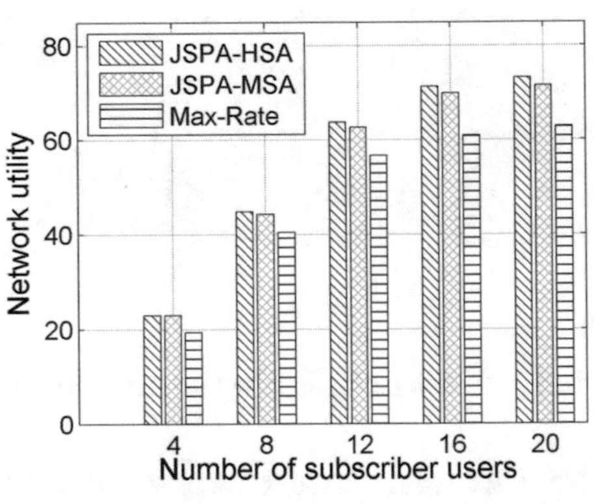

Fig. 6.5 Network utility versus number of subscriber users

backhaul inks. The backhaul link, cache storage, and BS bandwidth are denoted by B (Mbps), S (contents), and W (MHz), respectively. The process of network slicing will be carried out as per the tenant demand. The tuple $(\alpha_n, \beta_n, \gamma_n)$ denotes the slice allocation to a tenant n and it represents the fraction of (W, S, B) while fulfilling the conditions of $\beta_n = \sum_{n\in\mathcal{N}} \gamma_n = \sum_{n\in\mathcal{N}} \alpha_n = \sum_{n\in\mathcal{N}} \beta_n = 1$. $\boldsymbol{\alpha}$ is used for representation of vector $(\alpha_n)_{n\in\mathcal{N}}$, similar is the usage of $\boldsymbol{\gamma}$ and $\boldsymbol{\beta}$. The

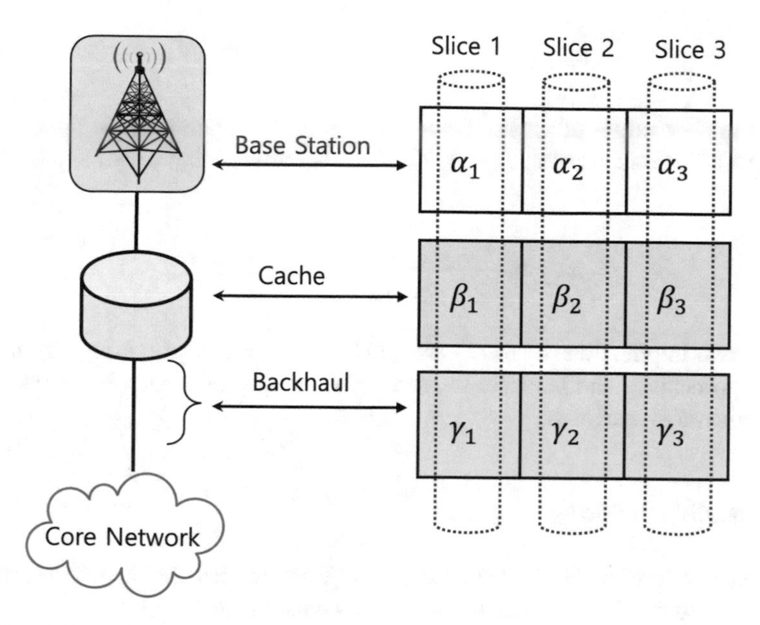

Fig. 6.6 A RAN network slicing model with three tenants

set $\mathcal{U}_n$ represents the users of tenant n. The set $\mathcal{F}$ denotes the set of equal size L contents. The pattern of the content access is modeled using distribution $p_{u_n,f}$ such that $\sum_{f\in\mathcal{F}} p_{u_n,f} = 1, \forall u_n \in \mathcal{U}_n$ for user $u_n \in \mathcal{U}_n$ of tenant n. A time slot model is used where decision of slicing is updated as per the tenants users demands. The content request rate of the tenant n's user is represented by $\lambda_{u_n}, \forall u_n \in \mathcal{U}_n$ for one time slot.

Cache Slicing

The cache hit probability of tenant n content f considering reset time-to-live (TTL) cache model is given by:

$$h_{n,f} = 1 - e^{-\sum_{u_n\in\mathcal{U}_n} \lambda_{u_n} F T_{n,f}}, \tag{6.22}$$

where $T_{n,f}$ denote the characteristic time of tenant n content f and it can be taken as timer in TTL caching. The number of tenant content must not be greater than the cache capacity, which is given by:

$$\sum_{f\in\mathcal{F}} h_{n,f} = \beta_n S, \quad \forall n \in \mathcal{N}. \tag{6.23}$$

Then each tenant n content request rate has smaller value due to large number of contents. Therefore, we can write:

$$h_{n,f} = \sum_{u_n \in \mathcal{U}_n} \lambda_{u_n} F T_{n,f}, \quad \forall n \in \mathcal{N}, \tag{6.24}$$

For smaller values of x, we have $e^{-x} = 1 - x$. The characteristic time of all the contents of tenants is same: $T_{n,f} = T_n, \forall f$. Therefore, using (6.24) and (6.23), we have:

$$h_{n,f} = \frac{\sum_{u_n \in \mathcal{U}_n} \lambda_{u_n} F}{\sum_{u_n \in \mathcal{U}_n} \lambda_{u_n}} \beta_n S, \quad \forall n \in \mathcal{N}. \tag{6.25}$$

The two implications of (6.25) are: (1) There is direct proportionality between the hit probability and request rate, and (2) when the cache size increases, the hit probability also increases.

BS—Bandwidth Slicing

The bandwidth of the BS for tenant n is $\alpha_n W$ when the BS slicing α_n is received by it. Then the downlink data rate of the tenant's n user is given by:

$$c_{u_n} = \alpha_n W \log(1 + \mathrm{SNR}_{u_n}), \quad \forall u_n \in \mathcal{U}_n, \tag{6.26}$$

where SNR_{u_n} is user u_n's SNR of the received signal. The SNR_{u_n} can be given by:

$$\mathrm{SNR}_{u_n} = \frac{g_{u_n} P_{u_n}}{\sigma_{u_n}^2 + I_{u_n}}, \tag{6.27}$$

where P_{u_n} represent the transmission power of the BS, $\sigma_{u_n}^2$ is the noise, I_{u_n} represents the other interference sources observed by user u_n, and g_{u_n} denote the channel gain. For user u_n, the missed rate and hit rate using (6.25) are given by:

$$\lambda_{u_n}^{hit} = \sum_f \lambda_{u_n} F h_{n,f} = \sum_f \lambda_{u_n} F \frac{\sum_{u_n \in \mathcal{U}_n} \lambda_{u_n} F}{\sum_{u_n \in \mathcal{U}_n} \lambda_{u_n}} \beta_n S \tag{6.28}$$

$$\lambda_{u_n}^{mis} = \lambda_{u_n} - \lambda_{u_n}^{hit}, \quad \forall u_n \in \mathcal{U}_n. \tag{6.29}$$

The tenant n base station slice serves the request of content f if it is found in storage of the cache. Therefore, the base station slice n load is given by:

$$\rho_n(\alpha_n, \beta_n) = \sum_{u_n \in \mathcal{U}_n} \frac{\lambda_{u_n}^{hit}}{\mu_{u_n}}, \tag{6.30}$$

where $\mu_{u_n} = c_{u_n}/L$ denote the user u_n's content service rate.

Backhaul Capacity Slicing

The requests that are missed are forwarded to content server through backhaul link. The tenant n delay cost of the backhaul link's given backhaul capacity slice $\gamma_n B$ is modeled according to M/M/1 queuing delay [4] and given by:

$$\Phi_n(\beta_n, \gamma_n) := \begin{cases} \frac{\lambda_n^{mis}}{\mu_n^{mis} - \lambda_n^{mis}}, & 0 \leq \lambda_n^{mis} < \mu_n^{mis}; \\ \infty, & \text{otherwise}, \end{cases} \tag{6.31}$$

where $\lambda_n^{mis} := \sum_{u_n \in \mathcal{U}_n} \lambda_{u_n}^{mis}$, and $\mu_n^{mis} = \gamma_n B / L$ denote the tenant n backhaul service rate.

6.3.2 RAN Network Slicing

This section presents the formulation of RAN slicing problem and derives a solutions for it.

Problem Formulation

The formulation of the RAN slicing is done as follows:

$$\min_{\alpha, \beta, \gamma \geq 0} \quad \sum_{n \in \mathcal{N}} \rho_n(\alpha_n, \beta_n) + \omega \sum_{n \in \mathcal{N}} \Phi_n(\beta_n, \gamma_n) \tag{6.32}$$

$$\text{s.t.} \quad \sum_{n \in \mathcal{N}} \alpha_n = \sum_{n \in \mathcal{N}} \beta_n = \sum_{n \in \mathcal{N}} \gamma_n = 1, \tag{6.33}$$

$$\rho_n(\alpha_n, \beta_n) \leq 1 - \epsilon, \qquad \forall n \in \mathcal{N}. \tag{6.34}$$

The objective of this problem deals with the minimization of two parts: (1) Cost of all slices backhaul links and (2) base station slices aggregated load. We see that

$$\rho_n(\alpha_n, \beta_n) = \sum_{u_n \in \mathcal{U}_n} \frac{\lambda_{u_n}^{hit}}{\mu_{u_n}} = \theta_n \frac{\beta_n}{\alpha_n}, \tag{6.35}$$

where $\theta_n := \sum_{u_n \in \mathcal{U}_n} \frac{\sum_{f \in \mathcal{F}} \lambda_{u_n} F \frac{\sum_{u_n \in \mathcal{U}_n} \lambda_{u_n} F}{\sum_{u_n \in \mathcal{U}_n} \lambda_{u_n}} S}{c_{u_n}/L}$.

We can rewrite the objective (6.32) as:

$$\sum_{n \in \mathcal{N}} \theta_n \frac{\beta_n}{\alpha_n} + \omega \sum_{n \in \mathcal{N}} \Phi_n(\beta_n, \gamma_n). \tag{6.36}$$

Equation (6.32) is non-convex, however biconvex problem. The problem has non-convex nature of all variables. It has convex nature with respect to γ_n and α_n, $\forall n \in \mathcal{N}$ given β_n, and vice versa. The two algorithms for solution of this biconvex problem are given in the next section.

6.3.3 Solution Approach

The centralized first algorithm has got inspiration from software defined network orchestration tools [8].

Centralized Approach

In this algorithm, the authors use the proximal block coordinate descent (PBCD) approach to solve a general multi-convex problem [11]. The controller computes solution of the following convex problem for getting $\boldsymbol{\beta}^{(k+1)}$ given $\boldsymbol{\gamma}^{(k)}$ and $\boldsymbol{\alpha}^{(k)}$ for each iteration k.

$$\min_{\boldsymbol{\beta}\geq 0} \quad \sum\nolimits_{n\in\mathcal{N}} \theta_n \frac{\beta_n}{\alpha_n^{(k)}} + \omega \sum\nolimits_{n\in\mathcal{N}} \Phi_n(\beta_n, \gamma_n^{(k)}) + \frac{\rho}{2}\|\beta - \beta^{(k)}\|^2 \tag{6.37}$$

$$\text{s.t.} \quad \sum\nolimits_{n\in\mathcal{N}} \beta_n = 1, \tag{6.38}$$

$$\theta_n \beta_n - (1-\epsilon)\alpha_n^{(k)} \leq 0, \quad \forall n \in \mathcal{N}. \tag{6.39}$$

The controller then updates $\boldsymbol{\alpha}^{(k+1)}$, $\boldsymbol{\gamma}^{(k+1)}$ based on $\boldsymbol{\beta}^{(k+1)}$ through solution of the convex problem given below:

$$\min_{\boldsymbol{\alpha},\boldsymbol{\gamma}\geq 0} \quad \sum\nolimits_{n\in\mathcal{N}} \theta_n \frac{\beta_n^{(k+1)}}{\alpha_n} + \omega \sum\nolimits_{n\in\mathcal{N}} \Phi_n(\beta_n^{(k+1)}, \gamma_n)$$

$$+ \frac{\rho}{2}\|\alpha - \alpha^{(k)}\|^2 + \frac{\rho}{2}\|\gamma - \gamma^{(k)}\|^2 \tag{6.40}$$

$$\text{s.t.} \quad \sum\nolimits_{n\in\mathcal{N}} \alpha_n = \sum\nolimits_{n\in\mathcal{N}} \gamma_n = 1, \tag{6.41}$$

$$\theta_n \beta_n^{(k+1)} - (1-\epsilon)\alpha_n \leq 0, \quad \forall n \in \mathcal{N}, \tag{6.42}$$

The operation of this algorithm is based on the alternative solution of (6.40) and (6.37) till convergence [11]. The controller (that can be realized by shared base station network manager [9]) first of all performs the collection of information of all tenants. The decentralized technique is imperative if the private information of the tenants is not available.

Decentralized Approach

The decentralized scheme is based on Jacobi-proximal alternating direction method of multipliers (JP-ADMM) [2]. The JP-ADMM is the variant of the ADMM and it converges faster than dual decomposition method [2]. Given aggregated values $\sum_{n\in\mathcal{N}} \beta_n^{(k)}$, $\sum_{n\in\mathcal{N}} \alpha_n^{(k)}$, $\sum_{n\in\mathcal{N}} \gamma_n^{(k)}$ and dual variables $(u_1^{(k)}, u_2^{(k)}, u_3^{(k)})$ of the previous iteration, at iteration k, every tenant n can find the solution of its problem as follows: Initially, tenant n performs updating of $\beta_n^{(k+1)}$ through solving

$$\begin{aligned} \min_{\beta_n\geq 0}. \quad & \theta_n \frac{\beta_n}{\alpha_n^{(k)}} + \omega\,\Phi_n(\beta_n, \gamma_n^{(k)}) + \frac{\tau_n}{2}(\beta_n - \beta_n^{(k)})^2 \\ & + \frac{\rho}{2}\Big(\beta_n + \sum\nolimits_{i\neq n} \beta_i^{(k)} - 1 - u_1^{(k)}/\rho\Big)^2 \\ \text{s.t.} \quad & \text{Constraint (6.39).} \end{aligned} \tag{6.43}$$

Then, with this $\beta_n^{(k+1)}$, the tenant n updates $(\alpha_n^{(k+1)}, \gamma_n^{(k+1)})$ by solving

$$\begin{aligned} \min_{\alpha_n,\gamma_n\geq 0}. \quad & \theta_n \frac{\beta_n^{(k+1)}}{\alpha_n} + \omega\,\Phi_n(\beta_n^{(k+1)}, \gamma_n) \\ & + \frac{\tau_n}{2}(\alpha_n - \alpha_n^{(k)})^2 + \frac{\rho}{2}\Big(\alpha_n + \sum\nolimits_{i\neq n} \alpha_i^{(k)} - 1 - \frac{u_2^{(k)}}{\rho}\Big)^2 \\ & + \frac{\tau_n}{2}(\gamma_n - \gamma_n^{(k)})^2 + \frac{\rho}{2}\Big(\gamma_n + \sum\nolimits_{i\neq n} \gamma_i^{(k)} - 1 - \frac{u_3^{(k)}}{\rho}\Big)^2 \\ \text{s.t.} \quad & \text{Constraint (6.42).} \end{aligned}$$

The controller updates the dual variables in the *second* step as follows:

$$u_i^{(k+1)} = u_i^{(k)} - \sigma\rho\Big(\sum\nolimits_{n=1}^{N} x_{i,n}^{(k+1)} - 1\Big), i = 1, 2, 3, \tag{6.44}$$

where $x_{i,n}^{(k+1)}$ represents $\gamma_n^{(k+1)}$, $\alpha_n^{(k+1)}$, $\beta_n^{(k+1)}$ for $i = 3, 2, 1$, respectively. The next step is to announce the dual variables along with aggregated variables to all tenants. The proposed schemes signaling can be implemented through type 5 [9], when σ and τ_n are selected as per [2]. The above-explained steps are performed repeatedly till the convergence is achieved.

6.3.4 Numerical Studies

In this section, an area of 500 m × 500 m is considered. The base station is deployed at the center with users having uniform distribution. Four tenants are considered and users within a tenant has a uniform distribution between [10, 30]. Apart from that all the users have same request rate. The bandwidth and transmission power of the base station are set to 20 MHz and 49 dBm, respectively. Apart from that, 50,000 contents each with size of 100 KB are considered and the popularity of content follows Zipf distribution and the value of ω is taken equal to 1.

Two scenarios (implemented in Julia) of four tenants and two tenants are considered, where backhaul capacity (Mbps) and cache capacity are set to 800 and 220 for two tenant case and 1200 and 220 for four tenant case, respectively (Table 6.3). It is illustrated in Fig. 6.7 that global optimal cost for tenant's is achieved by the proposed algorithms, although they do not exhibit theoretical global optimum for the problem (6.32).

The exhaustive search does not seem practical for the scenario of four tenants. Therefore, a comparison of convergence total cost between the proposed and IpOpt is made in lower plot of Fig. 6.7. Figure 6.7 reveals same convergence values for all schemes.

Figure 6.8 illustrates the results of 300 simulations which reveal that JP-ADMM scheme has slower convergence than PBCD. Apart from this, Fig. 6.9 shows the trade-off between backhaul cost and tenant load for scenario of four tenants through the turning parameter ω. Higher backhaul cost and lower tenant load is observed for smaller values of ω and vice versa. Figure 6.10 illustrates the convergence of backhaul slicing, cache, and BS. It is observed that higher demands tenants 4 and 1

Table 6.3 Capacity setting

Capacity	Two tenants	Four tenants
Cache (# Files)	220	220
Backhaul (Mbps)	800	1200

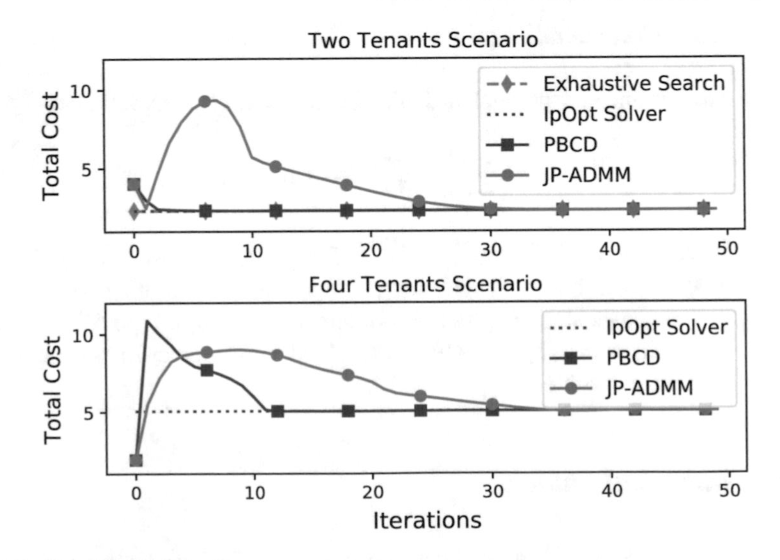

Fig. 6.7 Cost convergence

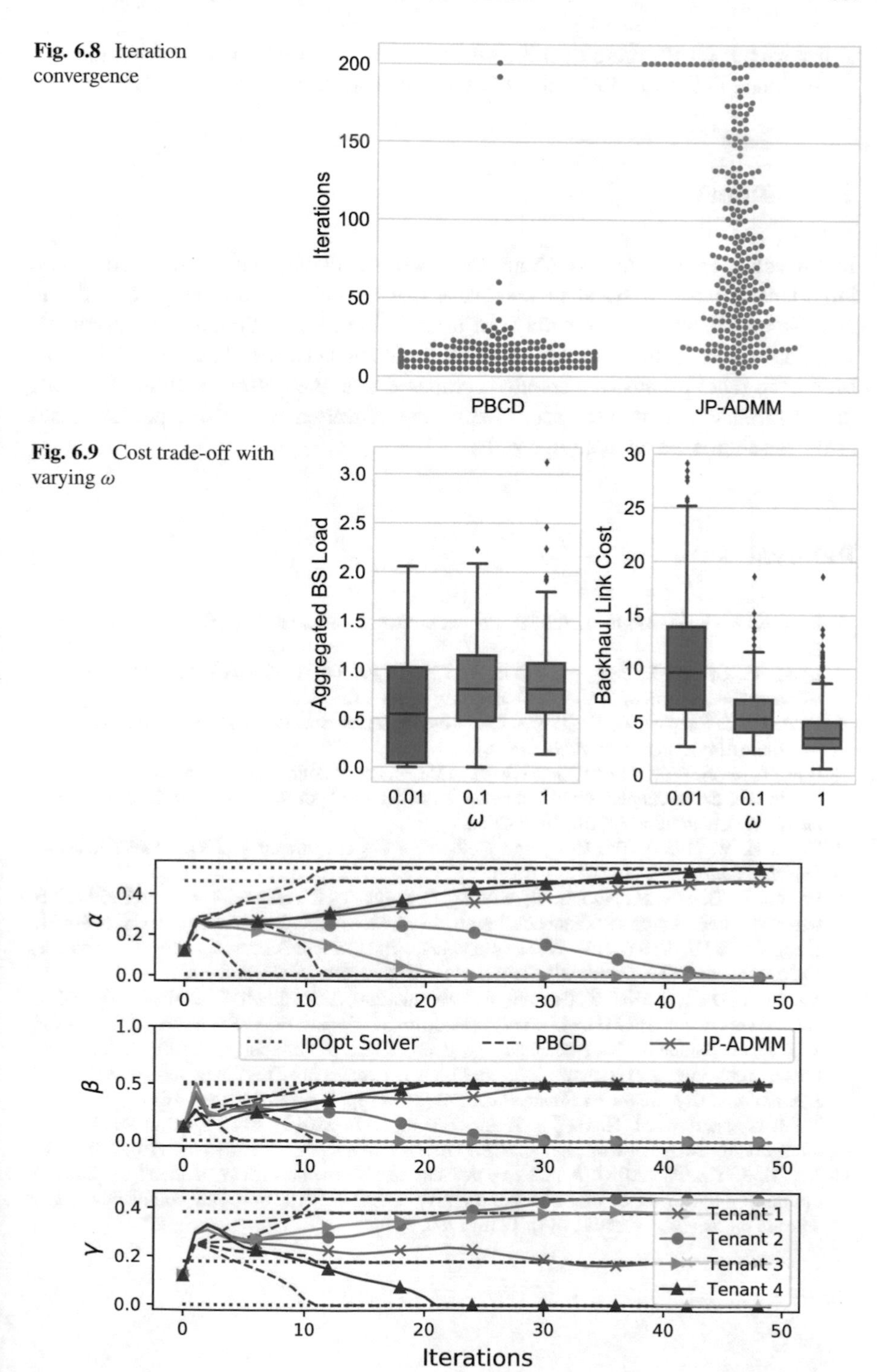

Fig. 6.8 Iteration convergence

Fig. 6.9 Cost trade-off with varying ω

Fig. 6.10 Slicing allocation of four tenants

utilize cache and BS more than the other tenants which get their contents through the backhaul link using 44% and 34% backhaul capacity.

6.4 Summary

In this chapter, we discuss about the network slicing while considering two important resources of backhaul and cache space which are becoming scarce due to massive proliferation of the number of users in the cellular system. Two proposals were presented in this chapter to create efficient network slices for end users. Moreover, these proposals were also compared with the optimal solution to depict the performance of the proposal. Finally, the convergence of these proposals has also been shown via simulation results.

References

1. Boyd, S., & Vandenberghe, L. (2004). *Convex optimization*. Cambridge: Cambridge University Press.
2. Deng, W., Lai, M.-J., Peng, Z., & Yin, W. (2017). Parallel multi-block ADMM with o(1/k) convergence. *Journal of Scientific Computing, 71*(2), 712–736.
3. Gale, D., & Shapley, L. S. (1962). College admissions and the stability of marriage. *The American Mathematical Monthly, 69*(1), 9–15.
4. Han, T., & Ansari, N. (2017). Network utility aware traffic load balancing in backhaul-constrained cache-enabled small cell networks with hybrid power supplies. *IEEE Transactions on Mobile Computing, 16*(10), 2819–2832.
5. Kuhn, H. W. (1955). The Hungarian method for the assignment problem. *Naval Research Logistics Quarterly, 2*(1–2), 83–97.
6. LeAnh, T., Tran, N. H., Ngo, D. T., & Hong, C. S. (2017). Resource allocation for virtualized wireless networks with backhaul constraints. *IEEE Communications Letters, 21*(1), 148–151.
7. Liang, C., & Yu, F. R. (2015). Wireless network virtualization: A survey, some research issues and challenges. *IEEE Communications Surveys & Tutorials, 17*(1), 358–380.
8. Nakao, A., Du, P., Kiriha, Y., Granelli, F., Gebremariam, A. A., Taleb, T., et al. (2017). End-to-end network slicing for 5G mobile networks. *Journal of Information Processing, 25*, 153–163.
9. Tseliou, G., Samdanis, K., Adelantado, F., Pérez, X. C., & Verikoukis, C. (2016). A capacity broker architecture and framework for multi-tenant support in LTE-A networks. In *2016 IEEE International Conference on Communications (ICC)* (pp. 1–6). Piscataway: IEEE.
10. Vo, P. L., Nguyen, M. N. H., Le, T. A., & Tran, N. H. (2018). Slicing the edge: Resource allocation for RAN network slicing. *IEEE Wireless Communications Letters, 7*(6), 970–973.
11. Xu, Y., & Yin, W. (2013). A block coordinate descent method for regularized multiconvex optimization with applications to nonnegative tensor factorization and completion. *SIAM Journal on Imaging Sciences, 6*(3), 1758–1789.

Chapter 7
Network Slicing: Dynamic Isolation Provisioning and Energy Efficiency

7.1 Introduction

In this chapter, we discuss about a novel resource allocation approach in which network's energy efficiency is maximized while providing the isolation of slicing. To address this problem, a hierarchical framework is proposed with three participants, i.e., cellular users, MVNOs, and InPs.

7.2 System Model and Problem

In this work [6], authors have considered two-tier network for downlink communication which consists of a single macro base station (MBS) and many small-cell base stations (SBSs) (Fig. 7.1). A single infrastructure provider (InP) is owning all of the MBS, SBSs, and spectrum resources to form a monopoly market. The InP is virtually serving the set of mobile virtual network operators (MVNOs) $\mathcal{M}$ with the help of individual contracts. However, each InP can provide services to its users which are different from the MVNOs users using the postpaid contracts. The users of MVNOs are assumed to get the services from a set $\mathcal{K}$ BSs having single MBS and containing $|\mathcal{K}| - 1$ SBSs. The services provided by an MVNO $m \in \mathcal{M}$ to a set $\mathcal{U}_m$ of signed up users with individual contracts are assumed to be different as compared to be served by different MVNOs. The set $\mathcal{S}_k$ denotes the set of users which have association with BS $k \in \mathcal{K}$ and the set $\mathcal{U} \triangleq \cup_{m=1}^{|\mathcal{M}|}\mathcal{U}_m = \cup_{k=1}^{|\mathcal{K}|}\mathcal{S}_k$ denotes all of the users. The set $\mathcal{C}$ of orthogonal sub-channels having bandwidth W is owned by InP. Co-channel deployment system is considered, which means that full frequency reuse [2] is utilized where all $|\mathcal{C}|$ sub-channels are available to be utilized by the users in a cell.

S. M. A. Kazmi et al., *Network Slicing for 5G and Beyond Networks*,
https://doi.org/10.1007/978-3-030-16170-5_7

7.2.1 Virtualization Model

Various options can be used for the isolation of the physical resource. One of the options is to use static sharing scheme where a fixed subset of physical resource is preassigned to each MVNO while restricting the access within the static subset. Another option is a generalized dynamic sharing scheme which does not restrict to use the resources, during assuring certain fixed demands or agreement of services (e.g., reducing the use of resource or throughput) [14] to accomplish the virtualization. This work has adopted the later virtualization scheme, where each InP uses a long-term data rate constraint to ensure agreement of services with MVNOs.

The virtualization and management of slices are executed by a wireless virtualization controller (or called a hypervisor in some existing studies) where the virtualization requirements (e.g., isolation, customization, and utilization) can be realized [4, 9]. The wireless virtualization controller can be implemented in a control center, centralized or distributed at each BS (Fig. 7.1).

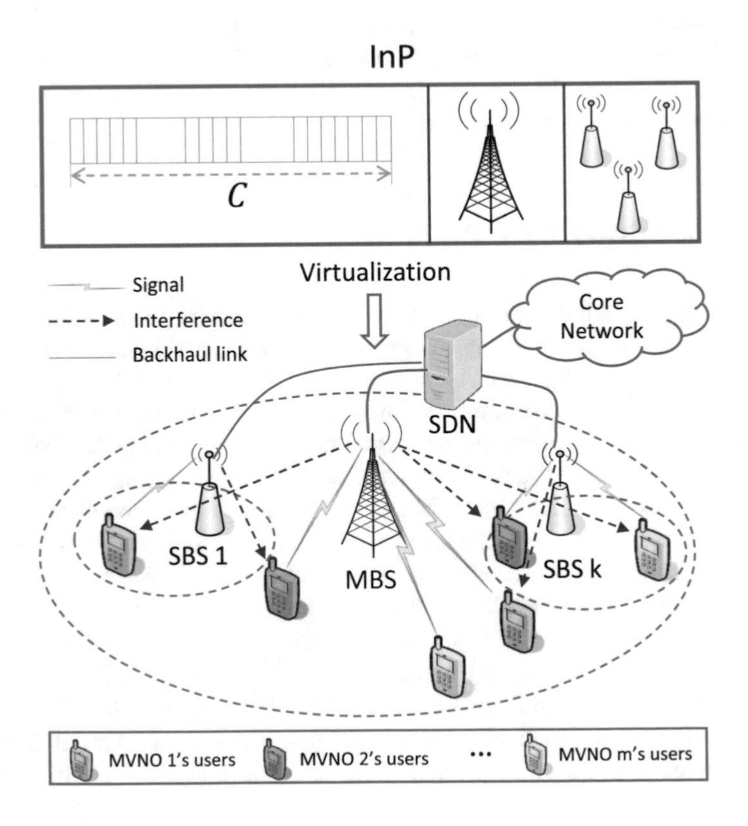

Fig. 7.1 System model: system model with InP owning the physical infrastructure and providing virtualized resources to multiple MVNOs users

In-Slice BS Allocation To characterize the BS allocation, let $\mathbf{A} \in \mathbb{R}^{|\mathcal{U}|\times|\mathcal{K}|}$ represent the matrix for all $|\mathcal{U}|$ cellular users associated with $|\mathcal{K}|$ BSs where the entries of the matrix are defined as follows:

$$a_{k,m,i} = \begin{cases} 1 & \text{if cellular user } i \text{ of MVNO } m \text{ is associated with BS} k \\ 0 & \text{otherwise.} \end{cases}$$

It is assumed that each cellular user is assigned only one BS, i.e.,

$$\sum_{k\in\mathcal{K}} a_{k,m,i} \leq 1, \ \forall i \in \mathcal{U}_m, \forall m \in \mathcal{M}. \tag{7.1}$$

In-Slice Sub-Channel Allocations To characterize the sub-channel assignments, let $\mathbf{Y} \in \mathbb{R}^{|\mathcal{U}|\times|\mathcal{C}|}$ represent the matrix for all $|\mathcal{U}|$ cellular users which are using $|\mathcal{C}|$ sub-channels where the entries of the matrix are given as follows:

$$y_{m,i}^{(c)} = \begin{cases} 1 & \text{if user } i \text{ of MVNO } m \text{ is allocated the sub-channel } c, \\ 0 & \text{otherwise.} \end{cases}$$

It is assumed that each user is allocated only one sub-channel in a BS:

$$\sum_{m_i\in\mathcal{S}_k} y_{m,i}^{(c)} \leq 1, \ \forall k \in \mathcal{K}, \forall c \in \mathcal{C}. \tag{7.2}$$

where m_i represents the cellular user i of MVNO m. In order to insure the agreement of services, at least one sub-channel is assigned to every user of an MVNO at any time period

$$\sum_{c\in\mathcal{C}} y_{m,i}^{(c)} \geq 1, \ \forall k \in \mathcal{K}, \forall m_i \in \mathcal{S}_k. \tag{7.3}$$

Transmission Rate Consider $h_{k,m,i}^{(c)}$ and $\sigma_{m,i}^{(c)}$ denote the channel power gain from BS k to the cellular user m_i and the noise power over sub-channel c, respectively. Denote $\mathbf{P}^{(c)} = \{P_0^{(c)}, P_1^{(c)}, \ldots, P_{|\mathcal{K}|}^{(c)}, \}$, $\mathbf{P}_k = \{P_k^{(1)}, P_k^{(2)}, \ldots, P_k^{(|\mathcal{C}|)}, \}$, and $\mathbf{P} = \{\mathbf{P}_0, \mathbf{P}_1, \ldots, \mathbf{P}_{|\mathcal{K}|}\}$ where $P_k^{(c)}$ is the transmit power of BS k on sub-channel c. As the co-channel deployment causes co-tier and cross-tier interference, the achievable signal-to-interference-and-noise ratio (SINR) for the communication among the cellular user m_i and BS k for the sub-channel c is given as

$$\Gamma_{k,m,i}^{(c)}(\mathbf{P}^{(c)}) = \frac{h_{k,m,i}^{(c)} P_k^{(c)}}{\sum_{l\in\mathcal{K},l\neq k} h_{l,m,i}^{(c)} P_l^{(c)} + \sigma_{m,i}^{(c)}}. \tag{7.4}$$

Considering that the Shannon's capacity is possible to be obtained, the throughput (in bps) for the cellular user m_i on sub-channel c from BS k is written as

$$R_{k,m,i}^{(c)}(\mathbf{P}^{(c)}) = W \ln(1 + \Gamma_{k,m,i}^{(c)}(\mathbf{P}^{(c)})). \tag{7.5}$$

Without any loss of generality, it is assumed that $W = 1$, unless stated otherwise.

Due to the limited backhaul capacity, the sum-rate of every BS k is restricted under its backhaul capacity, i.e.,

$$\sum_{m_i \in \mathcal{U}} \sum_{c \in \mathcal{C}} a_{k,m,i} y_{m,i}^{(c)} R_{k,m,i}^{(c)}(\mathbf{P}^{(c)}) \leq B_k, \ \forall k \in \mathcal{K}. \tag{7.6}$$

Energy Efficiency The aim of InP is to optimize the energy efficiency of the associated users for each serving MVNO in order to preserve operational cost like monetary payment for electric consumption. The performance metric for energy efficiency (bits/Hz per Joule) of all BSs is characterized as the proportion of the total data rate and the total energy consumed

$$\eta_{\mathrm{EE}} = \frac{\mathcal{R}(\mathbf{A}, \mathbf{Y}, \mathbf{P})}{\mathcal{P}(\mathbf{A}, \mathbf{Y}, \mathbf{P})} = \frac{\sum_{k \in \mathcal{K}} \sum_{m_i \in \mathcal{U}} \sum_{c \in \mathcal{C}} a_{k,m,i} y_{m,i}^{(c)} R_{k,m,i}^{(c)}(\mathbf{P}^{(c)})}{\sum_{k \in \mathcal{K}} \left(\sum_{m_i \in \mathcal{U}} \sum_{c \in \mathcal{C}} a_{k,m,i} y_{m,i}^{(c)} P_k^{(c)} + P_k^0 \right)}, \tag{7.7}$$

where P_k^0 represents the added circuit power consumed by BS k for the communication, which is considered to not depend on the data transmission power.

The InP creates virtual resource (VR) (i.e., slices) based on the request from MVNOs, and operates the slices and assigns them to the MVNOs subscribed users [9]. In general, the isolation between the slices of different MVNOs is required by any WNV scheme. In this work, a dynamic isolation scheme is assumed with no restriction on the assigned resource to the users of different MVNOs, which guarantees the isolation by achieving a certain contract agreement.

7.2.2 *MVNO Model*

By using the isolation, the users are scheduled by each MVNO and the data rate of the necessary users is found by the QoS demand of each MVNO. The goal of each MVNO is to boost the performance of its cellular users while minimizing cost paid to the InP by optimally choosing its requirements against the offered price of InP. Therefore, each MVNO is playing the role of broker for its cellular users by submitting their rate demands to the InP. The formulated problem for each MVNO m at the time instant t can be expressed as:

$$\max_{\mathbf{r}_m(t)} \sum_{i \in \mathcal{U}_m} \log(1 + r_{m,i}(t))$$
$$\text{s.t. } r_{m,i}(t) \geq r_{m,i}^{\min}, \ \forall i \in \mathcal{U}_m, \tag{7.8}$$
$$\beta_m(t) \sum_{i \in \mathcal{U}_m} r_{m,i}(t) \leq \mathfrak{B}_m^{\max},$$

where $\beta_m(t)$ denotes the price per unit of bandwidth (unit/bps) at time t for the MVNO m, $\mathfrak{B}_m^{\max}$ is the optimal cost for MVNO m, and $r_{m,i}(t)$ is the rate of cellular user $i \in \mathcal{U}_m$ on the time instant t.

As (7.8) is a convex problem, which can be solved for the MVNO as follows.

Lemma 7.1 *If the price from the InP is known, problem* (7.8) *can be solved for the MVNO as:*

$$r_{m,i}^*(t) = \left[\frac{\mathfrak{B}_m^{\max}}{|\mathcal{U}_m| \beta_m(t)} \right]_{r_{m,i}^{\min}}, \tag{7.9}$$

where $[\cdot]_x = \max\{\cdot, x\}$. *When the lower price is offered, the cellular users of MVNOs are offered with higher data rates. Conversely, when offered price is large, the rate is reduced while meeting the lower bound of the data rate* $r_{m,i}^{\min}$.

Multi-MVNO Model

In the competitive environment of multiple MVNOs hiring the resources from the InP, an interaction among MVOS is formulated by defining a non-cooperative game among MVOS for selfish contention of InP resources.

In every time slot t, it is considered that price offered by InP is an increasing convex function of the total rate requirements of all the MVNOs $r^{tot}(t) = \sum_m r_m(t)$, where $r_m(t)$ denotes the rate requirement from MVNO m in time slot t. The price given by InP is determined by $\beta(t) = \gamma(t) + \theta(t) r^{tot}(t)$, where $\gamma(t)$ is the wholesale price and $\theta(t)$ is a nonnegative coefficient with the unit of \$/bps. The value of $\gamma(t)$ and $\theta(t)$ is computed by InP based on the cost of serving MVNOs users. Given the price offered by InP $\beta(t)$, the MVNOs perform contention together in a rational way such the individual utility of each MVNO is maximized through a non-cooperative leasing game (NLG) defined as follows.

Definition 7.1 A non-cooperative leasing game $\mathcal{G}$ is defined as a triple: $\mathcal{G} \triangleq \{\mathcal{M}, (r_m)_{m \in \mathcal{M}}, (U_m)_{m \in \mathcal{M}}\}$, where $\mathcal{M}$ is the player set (set of MVNOs), $(r_m)_{m \in \mathcal{M}}$ is the strategy set, and $(U_m)_{m \in \mathcal{M}}$ is the payoff function.

Without any loss of generality, it is assumed that there is homogeneous data rate requirements $r_m^{\min}$ from the users of the same MVNO. Therefore, the payoff function of each MVNO m at each time slot t can be formulated as follows:

$$U_m(t) = (\alpha_m + \delta_m) r_m(t) - \beta(t) r_m(t), \tag{7.10}$$

where $\beta(t) = \gamma(t) + \theta(t) r^{tot}(t)$ is the price offered from InP, α_m is the retail price set by MVNO m, and δ_m is the saving cost of MVNO m (the cost saving from MVNO m because it is not paying to build the network, e.g., electric bill). The payoff function is composed of two components consisting of the profit of MVNO by providing the resources to the users and the cost of hiring the physical resources from InP to provide such services.

Theorem 7.1 *There exists a unique Nash equilibrium* $\mathbf{r}^{NE} = (r_1^{NE}, r_2^{NE}, \ldots, r_M^{NE})$ *in the proposed NLG game, where*

$$r_m^{\text{NE}} = \Bigg[\frac{1}{\theta}\Bigg(\alpha_m + \delta_m - \gamma - \frac{1}{M+1}\Bigg(\sum_{n \in \mathcal{M}} (\alpha_n + \delta_n) - M\gamma\Bigg)\Bigg)\Bigg]_{r_m^{\text{min}}}, \quad \forall m \in \mathcal{M}. \tag{7.11}$$

Proof The proof is provided in Appendix A of the supplementary material of [6].

Remark 7.1 It is observed that (1) different NE is produced as a result of having distinct values of γ and θ; (2) unique NE is obtained for the fixed values of γ and θ; (3) according to the retail price α_m and saving cost δ_m, each MVNO has corresponding strategies at the NE; and (4) the global knowledge about the total price and amount charged to all other MVNOs is to be known for an MVNO so that its strategy can be updated.

7.2.3 InP Model

A discrete queueing model of a single InP is considered to serve the set of MVNOs users. Consider that each agreement of services (isolation constraint) for MVNOs users is maintaining virtual queues denoted by $\mathbf{Q}(t) \triangleq \{Q_{m,i}(t), \forall m, i\}$, where $Q_{m,i}(t)$ represents the virtual queue backlog for user i of MVNO m at time slot t. The data rate requirement from MVNOs users $r_{m,i}(t)$ is regarded as the "arrival rate" and it is calculated from either (7.9) or (7.11) corresponding to independent MVNOs or competing MVNOs, respectively. The instantaneous rate of user i of MVNO m at time t is defined as $R_{m,i}(t) \triangleq \sum_{k \in \mathcal{K}} \sum_{c \in \mathcal{C}} a_{k,m,i} y_{m,i}^{(c)} R_{k,m,i}^{(c)}(\mathbf{P}^{(c)})(t)$. $R_{m,i}(t)$ is supposed to be the "service rate."İ Therefore, the evolution agreement of services queue is described as follows:

$$Q_{m,i}(t+1) = \max\{Q_{m,i}(t) - R_{m,i}(t) + r_{m,i}(t), 0\}. \tag{7.12}$$

In the long run, the development of a control algorithm in order to balance the contract agreement queues can be used to meet the data rate demands of individual user. Further details and explanations about the design of the algorithm for virtual queue can be found in [10] and the references therein. It can also be seen that discrete value is used to represent the time index t in the preceding mathematical statements which support that the proposed underlying queues are according to the discrete queuing model.

The corresponding long-term problem for InP with constraints on the balancing of the queue (i.e., the achieved data rate approaches the desired rate) can be expressed by:

$$\max_{(\boldsymbol{\beta},\mathbf{A},\mathbf{Y},\mathbf{P})} \limsup_{T\to\infty} \frac{1}{T}\sum_{t=0}^{T-1} \mathbb{E}\left\{\sum_{m\in\mathcal{M}}\sum_{i\in\mathcal{U}_m} \beta_m(t) r_{m,i}(t) + \omega\eta_{\mathrm{EE}}(t)\right\}$$

subject to:

$$\begin{aligned}
&C_1: \sum_{k\in\mathcal{K}} a_{k,m,i}(t) \le 1,\ \forall i\in\mathcal{U}_m, \forall m\in\mathcal{M},\\
&C_2: \sum_{m_i\in\mathcal{S}_k} y_{m,i}^{(c)}(t) \le 1,\ \forall k\in\mathcal{K}, \forall c\in\mathcal{C},\\
&C_3: \sum_{c\in\mathcal{C}} y_{m,i}^{(c)}(t) \ge 1,\ \forall k\in\mathcal{K}, \forall m_i\in\mathcal{S}_k,\\
&C_4: \limsup_{T\to\infty}\frac{1}{T}\sum_{t=0}^{T-1}\mathbb{E}\{Q_{m,i}(t)\} < \infty,\ \forall i,m,\\
&C_5: \sum_{m_i\in\mathcal{U}}\sum_{c\in\mathcal{C}} a_{k,m,i}(t) y_{m,i}^{(c)}(t) R_{k,m,i}^{(c)}(\mathbf{P}^{(c)}(t)) \le B_k,\ \forall k\in\mathcal{K},\\
&C_6: \sum_{m_i\in\mathcal{S}_k}\sum_{c\in\mathcal{C}} y_{m,i}^{(c)}(t) P_k^{(c)}(t) \le P_k^{\max},\ \forall k\in\mathcal{K},\\
&C_7: a_{k,m,i}(t), y_{m,i}^{(c)}(t)\in\{0,1\}, P_k^{(c)}(t)\ge 0,\ \forall i,m,k,c,\\
&C_8: \beta^{\min} \le \beta_m(t) \le \beta^{\max},\ \forall m,
\end{aligned} \tag{7.13}$$

where ω denotes the prices on a unit power efficiency. The aim of every InP is to boost its profit of the MVNOs users, i.e., the first expression, and to achieve the optimal power efficiency, i.e., the later expression. C_4 allows to ensure the isolation provisioning. Constraint C_5 denotes the limitations on the backhaul capacity of BS. Hence, the problem statement of InP is expressed as: for the dynamic network designed by (7.12), developing a protocol that selects the appropriate cost offered to the MVNOs, allocation of physical resource for the associated users of MVNOs in order to boost the mean time profit and power efficiency while satisfying other balance constraints of the isolation queues.

7.3 Solution Approach Based on Lyapunov Online Algorithm

As a general stochastic problem is denoted by (7.13), Lyapunov optimization techniques in [10] are used to develop an online algorithm in order to transform the average-based optimization problem into the single time slot optimization problem, such that the solution can be obtained by using the instantaneous channel state information (CSI), queue state information (QSI), and arrival rate of each user, which can significantly reduce the computation complexity.

To achieve this goal, the following Lyapunov function is considered

$$L(\mathbf{Q}(t)) \triangleq \frac{1}{2} \sum_{m \in \mathcal{M}} \sum_{i \in \mathcal{U}_m} \left(Q_{m,i}(t)\right)^2. \tag{7.14}$$

The intuition of reducing $L(\mathbf{Q}(t))$ by taking appropriate actions may proceed to preserve the balance of all queues, which results in the guarantee of balancing the rates of long-term users with the allocated rates (i.e., the constraints in the agreement are guaranteed). The queuing state of the system at time t is given as $\mathbf{Q}(t) \triangleq \{Q_{m,i}(t), \forall m, i\}$. The one-slot conditional Lyapunov drift is defined as

$$\Delta(\mathbf{Q}(t)) \triangleq \mathbb{E}\{L(\mathbf{Q}(t+1)) - L(\mathbf{Q}(t)) | \mathbf{Q}(t)\}. \tag{7.15}$$

The goal of this algorithm is to reduce the upper bound on the following *drift-plus-penalty* expression [10]:

$$\begin{aligned} &\Delta_V(\mathbf{Q}(t)) \triangleq \\ &\Delta(\mathbf{Q}(t)) - V\mathbb{E}\left\{ \sum_{m \in \mathcal{M}} \sum_{i \in \mathcal{U}_m} \beta_m(t) r_{m,i}(t) + \omega \eta_{\mathrm{EE}}(t) | \mathbf{Q}(t) \right\}, \end{aligned} \tag{7.16}$$

where the performance bound of the algorithm is controlled by nonnegative weight V. From (7.12) and (7.14)

Lemma 7.2 *For any feasible control action under constraints C_1, C_2, C_3, C_5, C_6, C_7, and C_8 that can be implemented at time slot t, we have the following inequality:*

$$\begin{aligned} \Delta_V(\mathbf{Q}(t)) \leq D + &\sum_{m \in \mathcal{M}} \sum_{i \in \mathcal{U}_m} Q_{m,i}(t) \left(r_{m,i}(t) - R_{m,i}(t)\right) \\ &- V\mathbb{E}\left\{ \sum_{m \in \mathcal{M}} \sum_{i \in \mathcal{U}_m} \beta_m(t) r_{m,i}(t) + \omega \eta_{EE}(t) | \mathbf{Q}(t) \right\}, \end{aligned} \tag{7.17}$$

where D is a constant given as

$$D \triangleq \frac{1}{2} \sum_{m \in \mathcal{M}} \sum_{i \in \mathcal{U}_m} \left(r_{m,i}^{\max 2} + R_{m,i}^{\max 2} \right), \tag{7.18}$$

where $R_{m,i}^{\max}$ and $r_{m,i}^{\max}$ are finite because of the limited transmission power in (7.5).

Proof The proof is provided in Appendix B of the supplementary material of [6].

An online approach is presented for the solution of (7.13), where the upper bound is reduced for the drift-plus-penalty expressed in Lemma 7.2, as given in Algorithm 1. This online approach is dependent on the drift-plus-penalty minimization algorithm, which works on the following principle: On each slot t, all queues $Q_{m,i}(t)$ are noted. After that, the InP find the solution of (7.19) such that $\sum_{m \in \mathcal{M}} \sum_{i \in \mathcal{U}_m} \left[V\beta_m(t) r_{m,i}(t) - (r_{m,i}(t) - R_{m,i}(t)) Q_{m,i}(t) \right] + V\omega\eta_{\text{EE}}(t)$ is maximized or its dual is minimized, i.e., right-hand side of (7.17). After that it is considered that $\omega = 1$.

7.3.1 Performance Bounds for the Online Algorithm

The given problem (7.19) is a nonlinear non-convex combinatorial problem, which is not easy to solve. In order to solve (7.19), the analysis of the performance of control actions is presented to find a local optimal inside the additive constant of the supremum. For that purpose, *C-additive approximation* [10] is defined in the following.

Definition 7.2 For the expressed constant $C \geq 0$, a *C-additive approximation* of the drift-plus-penalty minimization protocol is the chosen action in every time slot which produces a conditional expected value on the right-hand side of the drift-plus-penalty (given $\mathbf{Q}(t)$) at time slot t which is under C from the supremum on the feasible set of actions.

On the basis of Definition 7.1, a local optimization protocol is developed which is introduced in the following sections.

Define $f(t) \triangleq \sum_{m \in \mathcal{M}} \sum_{i \in \mathcal{U}_m} \beta_m(t) r_{m,i}(t) + \eta_{\text{EE}}(t)$, $f^{\max} \triangleq \max f(t)$, and $f^* \triangleq f(\boldsymbol{\beta}^*, \mathbf{A}^*, \mathbf{Y}^*, \mathbf{P}^*)$ where $(\boldsymbol{\beta}^*, \mathbf{A}^*, \mathbf{Y}^*, \mathbf{P}^*)$ is a theoretical optimal solution of (7.19) [10]. From the Lyapunov optimization approach, the following result is obtained.

Theorem 7.2 *For the positive values of the constants D, V, C, and ϵ such that for all time slots t and all possible values of $\mathbf{Q}(t)$, the Lyapunov drift meets the following requirement:*

Algorithm 1 Online algorithm to solve problem (7.13)

1: Initialization: $V > 0$, $Q_{m,i}(0) \leftarrow 0, \forall m, i$, and $t \leftarrow 0$;
2: **loop**
3: Collect the current QSI $Q_{m,i}(t)$;
4: Collect the demand request from MVNOs' users $r_{m,i}(t)$;
5: Solve the following optimization problem:

$$\begin{aligned}
&\max_{(\boldsymbol{\beta},\mathbf{A},\mathbf{Y},\mathbf{P})} \sum_{m\in\mathcal{M}}\sum_{i\in\mathcal{U}_m}[V\beta_m(t)r_{m,i}(t) \\
&\qquad - (r_{m,i}(t) - R_{m,i}(t))Q_{m,i}(t)] + V\omega\eta_{\mathrm{EE}}(t) \\
&\text{subject to:} \\
&C_1: \sum_{k\in\mathcal{K}} a_{k,m,i}(t) \le 1,\ \forall i \in \mathcal{U}_m, \forall m \in \mathcal{M}, \\
&C_2: \sum_{m_i\in\mathcal{S}_k} y^{(c)}_{m,i}(t) \le 1,\ \forall k \in \mathcal{K}, \forall c \in \mathcal{C}, \\
&C_3: \sum_{c\in\mathcal{C}} y^{(c)}_{m,i}(t) \ge 1,\ \forall k \in \mathcal{K}, \forall m_i \in \mathcal{S}_k, \\
&C_5: \sum_{m_i\in\mathcal{U}}\sum_{c\in\mathcal{C}} a_{k,m,i}(t)y^{(c)}_{m,i}(t)R^{(c)}_{k,m,i}(\mathbf{P}^{(c)}(t)) \le B_k,\ \forall k, \\
&C_6: \sum_{m_i\in\mathcal{S}_k}\sum_{c\in\mathcal{C}} y^{(c)}_{m,i}(t)P^{(c)}_k(t) \le P^{\max}_k,\ \forall k \in \mathcal{K}, \\
&C_7: a_{k,m,i}(t),\ y^{(c)}_{m,i}(t) \in \{0,1\},\ P^{(c)}_k(t) \ge 0,\ \forall i, m, k, c, \\
&C_8: \beta^{\min} \le \beta_m(t) \le \beta^{\max},\ \forall m;
\end{aligned} \tag{7.19}$$

6: Update queues states $Q_{m,i}(t+1)$;
7: $t \leftarrow t+1$;
8: **end loop**

$$\begin{aligned}
\Delta(\boldsymbol{Q}(t)) - V\mathbb{E}\{f(t)|\boldsymbol{Q}(t)\} \le D + C& \\
- \epsilon \sum_{m\in\mathcal{M}}\sum_{i\in\mathcal{U}_m} Q_{m,i}(t) - Vf^*&,
\end{aligned} \tag{7.20}$$

then time average utility and virtual queue length meet this requirement

$$\limsup_{T\to\infty} \frac{1}{T}\sum_{t=0}^{T-1}\sum_{m\in\mathcal{M}}\sum_{i\in\mathcal{U}_m} \mathbb{E}\{Q_{m,i}(t)\} \le \frac{D + C + V(f^{\max} - f^*)}{\epsilon}, \tag{7.21}$$

$$\liminf_{t\to\infty} \frac{1}{T}\sum_{t=0}^{T-1} \mathbb{E}\{f(t)\} \ge f^* - \frac{D+C}{V}, \tag{7.22}$$

where C and ϵ are obtained from the performance gap between the optimal solution and the sub-optimal solution obtained by the proposed algorithmic framework.

Remark 7.2

1. Getting the globally optimal solution to (7.19) while using the polynomial time algorithm is extremely difficult and near to impossible. Therefore, it is better to focus on designing a relatively easy and low-complexity algorithms which results in optimal solutions to (7.19) as compared to the globally optimal solution. In general, it is difficult to quantify the gap C that the algorithms can achieve.
2. It is shown in Theorem 7.2 that an $[\mathcal{O}(1/V), \mathcal{O}(V)]$ trade-off between the average revenue, average energy efficiency, and virtual queue backlogs can be obtained using the proposed online algorithm [10]. The achievable performance can be improved by increasing the control parameter V, at the cost of facing the higher queuing delay. Therefore, it is necessary to select the appropriate value of V in order to achieve the desired performance and isolation provisioning in realistic WNV.

7.3.2 Decomposition Approach

At the start, the data rate demands of each MVNO are computed from (7.9) or (7.11) using the price offered by InP $\beta(t)$ and the demands are forwarded to the InP. After that, InP runs the price decision algorithm using these data rate demands in order to compute the price charged to MVNOs as shown in Sect. 7.3.6. Then various low time-scale algorithms are used to perform the BS assignment and resource allocation by using the computed prices from the pricing decision algorithm. The next step after BS assignment to all of the subscribed users of MVNOs as described in Sect. 7.3.4 using the BS assignment algorithm, the distributed algorithm to jointly optimize the sub-channel allocation and energy efficiency is proceeded in order to perform the sub-channel resources to the users and energy efficiency to the InP by individual BS as described in Sect. 7.3.5. Such design helps to tackle the problem in (7.19) by making it analytically tractable.

7.3.3 Stackelberg Game and Stackelberg Equilibrium

For the cooperation among InP and MVONs, a two-level Stackelberg game is defined. The first stage allows the InP to offer the suitable prices in order to maximize its profit which is proceeded by the second stage where the MVNOs set the data rate requirements of their subscribed users by taking into account the price offered by the InP. A simple one-on-one interaction is performed when there are independent MVNOs having no interaction among other MVNOs.

Suppose $U_{\text{InP}}(\boldsymbol{\beta}, \mathbf{A}, \mathbf{Y}, \mathbf{P}, \mathbf{r})$ represents the utility of the InP, and $U_m(\mathbf{r}_m, \beta_m)$ represents the utility of MVNO m. The solution of InPs strategy is represented by $(\boldsymbol{\beta}^*, \mathbf{A}^*, \mathbf{Y}^*, \mathbf{P}^*)$ and the solution for the MVNOs strategy by $(\mathbf{r}^*)$, we have the following definition.

Definition 7.3 $(\boldsymbol{\beta}^*, \mathbf{A}^*, \mathbf{Y}^*, \mathbf{P}^*, \mathbf{r}^*)$ is a Stackelberg equilibrium for a Stackelberg game if it satisfies the following conditions for any values of $(\boldsymbol{\beta}, \mathbf{A}, \mathbf{Y}, \mathbf{P}, \mathbf{r})$:

$$U_{\text{InP}}(\boldsymbol{\beta}^*, \mathbf{A}^*, \mathbf{Y}^*, \mathbf{P}^*, \mathbf{r}^*) \geq U_{\text{InP}}(\boldsymbol{\beta}, \mathbf{A}, \mathbf{Y}, \mathbf{P}, \mathbf{r}^*), \ \forall \boldsymbol{\beta}, \mathbf{A}, \mathbf{Y}, \mathbf{P},$$

$$U_m(\mathbf{r}_m^*, \beta_m^*) \geq U_m(\mathbf{r}_m, \beta_m^*), \ \forall r_{m,i}.$$

The explanation of the Stackelberg equilibrium of the game defined in this paper is given as follows: As a first step, the problem of MVNOs is solved using the fixed price β by applying the non-cooperative game or independently. After that, the data rate demands $(\mathbf{r}^*)$ or $(\mathbf{r}^{\text{NE}})$ computed by solving the MVNOs problem are used by the InP to solve its problem of finding the optimal strategy $(\boldsymbol{\beta}^*, \mathbf{A}^*, \mathbf{Y}^*, \mathbf{P}^*)$. As a result of convex optimization problem for the MVNOs, it can satisfy the second inequality in Definition 7.1. On the other hand, the non-convexity of the InP problem for the short-time scale makes it near to impossible to find the globally optimal solution for the InPs strategy. Therefore more than one possible SE can be obtained locally for the defined game.

7.3.4 Base Station Assignment

After the assumption of equal division of the total power of BS k among all of the sub-channels, each UE m_i becomes similar for all the channels allocated by the BS. As a result, it is possible to eliminate the variable of sub-channel assignment in the problem of BS assignment. Therefore the BS allocation problem can be expressed in the following:

$$\max_{(\mathbf{A})} \ U_{\text{InP}} = V \frac{\sum\limits_{k\in\mathcal{K}} \sum\limits_{m_i\in\mathcal{U}} a_{k,m,i} R_{k,m,i}(\mathbf{P})}{\sum\limits_{k\in\mathcal{K}} \left(\sum\limits_{m_i\in\mathcal{U}} a_{k,m,i} P_k + P_k^0 \right)} + \sum_{m_i\in\mathcal{S}_k} \sum_{k\in\mathcal{K}} Q_{m,i} a_{k,m,i} R_{k,m,i}(\mathbf{P}) \tag{7.23}$$

$$\text{s.t. } C_1,$$

$$C_5' : \sum_{m_i\in\mathcal{U}} a_{k,m,i} R_{k,m,i}(\mathbf{P}) \leq B_k, \ \forall k \in \mathcal{K},$$

$$C_7' : a_{k,m,i} \in \{0, 1\}, \ \forall i, m, k,$$

where $P_k = P_k^{\max}/|\mathcal{C}|$. Since it can be seen that the above optimization problem in (7.23) is yet a combinatorial problem; however, as the BS assignment is the only binary variable in (7.23), the problem is possible to be modeled as a matching problem [13]. Therefore, a distributed algorithm is developed using the matching game to solve the optimization problem (7.23)[1] [5, 13] as given in the next section.

BS Assignment scheme

A two-sided matching game can be used to model problem (7.23) by considering two disjoint sets of agents, the set of MVNO users, $\mathcal{U}$, and the set of BSs, $\mathcal{K}$. Only a single BS k is assigned to each user i of MVNO m in the defined game. On the other hand, multiple number of users can be supported by a BS k based on the backhaul capacity B_k. The proposed design corresponds to a *one-to-many matching* given by the tuple $(\mathcal{U}, \mathcal{K}, B_k, \succ_{\mathcal{U}}, \succ_{\mathcal{K}})$. Here, $\succ_{\mathcal{U}} \triangleq \{\succ_{m_i}\}_{m_i \in \mathcal{U}_m}$ and $\succ_{\mathcal{K}} \triangleq \{\succ_k\}_{k \in \mathcal{K}}$ denote the set of the preference relations of the MVNOs users and BSs, respectively.

Definition 7.4 A *matching* μ is defined by a function from the set $\mathcal{U} \cup \mathcal{K}$ into the set of elements of $\mathcal{U} \cup \mathcal{K}$ such that:

1. $|\mu(m_i)| \leq 1$ and $\mu(m_i) \in \mathcal{K}$,
2. $|\mu(k)| \leq \bar{\mathcal{U}}_k$ and $\mu(k) \in 2^{|\mathcal{U}|} \cup \phi$,
3. $\mu(m_i) = k$ if and only if m_i is in $\mu(k)$,

where $\bar{\mathcal{U}}_k$ represents the dynamic quota of a BS k (i.e., number of subscribed users to BS k) such that constraint $C_5^{'}$ is ensured, and $|\mu(.)|$ represents the cardinality of the matching outcome $\mu(.)$.

Preference Function of Agents

The preference profile of each user m_i of MVNO is built by computing the data rate of individual user m_i for each BS k which is then sorted in the descending order. Finally the preference profile $\mathcal{P}_{m_i}$ denotes the vector containing the utilities of every MVNO user m_i, as follows:

$$U_{m_i}(k) = \left[R_{k,m,i}(\mathbf{P})\right]_{k \in \mathcal{K}}. \tag{7.24}$$

Here, every user m_i of MVNO wants to get associated with a BS k so that it can increase its maximum utility $U_{m_i}(\mathcal{K})$. $k \succ_{m_i} k'$ is used to refer to the MVNO user m_i preference of BS k over BS k', i.e., $U_{m_i}(k) > U_{m_i}(k')$.

To implement the distributed algorithm, the optimization problem of energy efficiency for all BSs is decoupled into individual BS utility function for each BS k as follows:

[1]Employing distributed algorithms for BS assignment enables to realize a scalable and low computation solution.

$$U_k(m_i) = \left[V \frac{R_{k,m,i}(\mathbf{P})}{P_k} + Q_{m,i} R_{k,m,i}(\mathbf{P}) \right]_{m_i \in \mathcal{U}}. \tag{7.25}$$

As given before, the preference profile $\mathcal{P}_k$ of BS k is maintained for the MVNO users m_i which is the backhaul capacity constrained energy efficiency maximization problem. The function given in (7.25) is used to compute the energy efficiency of BS k for every user which is arranged in the decreasing order in the preference profile $\mathcal{P}_k$. It can be seen that, in the defined game, the quota of BS k (i.e., $\bar{\mathcal{U}}_k$) is dynamic as compared to the traditional one-to-many matching games [13]. Therefore the proposed game deploys a dynamic quota $\bar{\mathcal{U}}_k$ as a BS k can serve a number of users (with heterogeneous rates, i.e., $R_{k,m,i}$ for user m_i) subject to the backhaul capacity constraint on that BS k, i.e., $C_5^{'}$. The achievable rate of each user m_i for a BS k is denoted by $\bar{\mathcal{U}}_k^{m_i} = R_{k,m,i}$.

As shown in the defined double-sided game in Definition 7.2, the objective is to look for a *stable matching*, which is a very important solution concept [13]. In order to develop a stable matching, there should not be any blocking pair. On the other hand, dynamic quota of BSs produces new challenges to refrain the utilization of the standard deferred-acceptance algorithm. For that purpose, the formal definition of the blocking pair is necessary for the formulated game as follows.

Definition 7.5 μ is considered as *stable* matching when no blocking pair (m_i, k) can be found, where $m_i \in \mathcal{U}, k \in \mathcal{K}$, such that $\bar{\mathcal{U}}_k^{res} \geq \bar{\mathcal{U}}_k^{m_i}$, $m_i \succ_k \emptyset$, and $k \succ_{m_i} \mu(m_i)$, where $\mu(m_i)$ denotes the selected matched peers of m_i.

Here, the remaining quota of BS K is denoted by $\bar{\mathcal{U}}_k^{res} = \bar{\mathcal{U}}_k - \sum_{m_i \in \mu(k)} \bar{\mathcal{U}}_k^{m_i}$. *The quota of BS $k \in \mathcal{K}$ is satisfied if $\bar{\mathcal{U}}_k^{res} < \bar{\mathcal{U}}_k^{m_i}$ for a user $m_i \in \mathcal{U}$.* Definition 7.3 is dependent on the given insight. If a BS k has sufficient quota $\bar{\mathcal{U}}_k^{res}$ to associate a cellular user m_i (i.e., $\bar{\mathcal{U}}_k^{res} \geq \bar{\mathcal{U}}_k^{m_i}$) and the user m_i is ready to get the proposal of k on the previously selected $\mu(m_i)$ (i.e., $k \succ_{m_i} \mu(m_i)$), then k and m_i can change the current selection of matching in order to create blocking pair. A matching cannot be changed when no blocking pairs are available.

Distributed Base Station Assignment Algorithm

To get the solution of this game, an algorithm is proposed for BS assignment in a distributed way by updating the deferred-acceptance algorithm in [13] in order to develop a stable algorithm for matching as given in Algorithm 2. First the algorithm is initialized, followed by building of preference profiles by both sides, i.e., $\mathcal{P}_k$ for BS k and $\mathcal{P}_{m_i}$ for user m_i. On every iteration t, every user m_i of an MVNO gets the preferences from BS k which are ranked m_i as the highest in $\mathcal{P}_k[t]$ having sufficient quota to accept the user m_i (lines 5–8). It is to be noted that every BS k has already defined some quota of $\bar{\mathcal{U}}_k$. Each BS k tries to suggest the proposals until the existence of unassigned users m_i of MVNO or the remaining quota $\bar{\mathcal{U}}_k$ is having enough resources to serve at least one more MVNO user m_i,

Algorithm 2 Distributed base station assignment algorithm

1: ***Phase 1: Initilization***:
2: **input**: $\mathcal{P}_k, \mathcal{P}_{m_i}, \forall m_i, k$;
3: **initialize**: $t = 0$, $\mu[0] \triangleq \{\mu[0](k), \mu[0](m_i)\}_{k\in\mathcal{K}, m_i\in\mathcal{U}_m} = \emptyset$, $\mathcal{L}_{m_i}[0] = \emptyset, \bar{\mathcal{U}}_k^{res}[0] = \bar{\mathcal{U}}_k$, $\mathcal{P}_k[0] = \mathcal{P}_k, \mathcal{P}_{m_i}[0] = \mathcal{P}_{m_i}, \forall k, m_i$;
4: ***Phase 2: Matching***:
5: **repeat**
6: $\quad t \leftarrow t + 1$;
7: $\quad$ **for** $k \in \mathcal{K}$, propose m_i according to $\mathcal{P}_k[t]$ **do**
8: $\quad\quad$ **while** $k \notin \mu[t](m_i)$ and $\bar{\mathcal{U}}_k^{res}[t] \geq \bar{\mathcal{U}}_k^{m_i}[t]$ **do**
9: $\quad\quad\quad$ **if** $k \succ_{m_i} \mu[t](m_i)$ **then**
10: $\quad\quad\quad\quad \mu[t](m_i) \leftarrow \mu[t](m_i) \setminus k'$;
11: $\quad\quad\quad\quad \bar{\mathcal{U}}_{k'}^{res}[t] \leftarrow \bar{\mathcal{U}}_{k'}^{res}[t] + \bar{\mathcal{U}}_{k'}^{m_i}[t]$;
12: $\quad\quad\quad\quad \mu[t](m_i) \leftarrow k$;
13: $\quad\quad\quad\quad \bar{\mathcal{U}}_k^{res}[t] \leftarrow \bar{\mathcal{U}}_k^{res}[t] - \bar{\mathcal{U}}_k^{m_i}[t]$;
14: $\quad\quad\quad\quad R_{m_i}[t] = \{k' \in \mathcal{P}_{m_i}[t] | k \succ_{m_i} k'\}$;
15: $\quad\quad\quad$ **else**
16: $\quad\quad\quad\quad R'_{m_i}[t] = \{k \in \mathcal{K} | \mu[t](m_i) \succ_{m_i} k\}$;
17: $\quad\quad\quad$ **end if**
18: $\quad\quad\quad \mathcal{L}_{m_i}[t] = \{R_{m_i}[t]\} \cup \{R'_{m_i}[t]\}$;
19: $\quad\quad\quad$ **for** $l \in \mathcal{L}_{m_i}[t]$ **do**
20: $\quad\quad\quad\quad \mathcal{P}_l[t] \leftarrow \mathcal{P}_l[t] \setminus \{m_i\}$;
21: $\quad\quad\quad\quad \mathcal{P}_{m_i}[t] \leftarrow \mathcal{P}_{m_i}[t] \setminus \{l\}$;
22: $\quad\quad\quad$ **end for**
23: $\quad\quad$ **end while**
24: $\quad$ **end for**
25: **until** $\mu[t] = \mu[t-1]$
26: ***Phase 3: BS Assignment***: $\mu \mapsto \mathbf{A}$;

i.e., $\bar{\mathcal{U}}_k^{res} \geq \bar{\mathcal{U}}_k^{m_i}$ (line 8). After getting the proposal from BS k, each m_i: (1) the proposal is accepted temporarily in case of k has preference over its current matching peer, i.e., $k \succ_{m_i} \mu[t](m_i)$ (line 9), and modifies its available matching $\mu[t](m_i)$ by replacing the previously accepted BS k' and updating the next proposal by BS k. Moreover, the remaining quota is also modified by both of the rejected BS k' and accepted BS k (lines 10–13). It can be seen that every BSs k' having lower rank as compared to the current matched BS k computes a set denoted by $R_{m_i}[t]$ (line 14); (2) conversely, it refuses the proposal of BS k and finds all of the low-rank BSs compared to the current matching partner $\mu[t](m_i)$ and put them in the set $R'_{m_i}[t]$ (lines 15–17). The BS having the lowest preference $k \in \mathcal{L}_{m_i}[t]$ is therefore rejected, and finally replaced from the preference list $\mathcal{P}_{m_i}[t]$, and in the same way these BSs are also replaced m_i from their corresponding preference list $\mathcal{P}_l[t]$ (lines 18–20). These steps are revised unless the matching procedure is converged, i.e., $\mu[t] = \mu[t-1]$ (line 21). It can be seen that by the method of BS proposing, the guarantee of matching no BS k to a user m_i having lower rank as compared to the current matched user set, i.e., $\mu(k)$ is ensured. In the final step of the BS assignment procedure, the outcome $\mu[t]$ is converted to a feasible BS assignment vector $\mathbf{A}$ of problem (7.23) (line 22), i.e., $\mu \mapsto \mathbf{A}$.

Property 1 Algorithm 2 converges to a stable assignment.

Proof The proof is provided in Appendix D of the supplementary material of [6].

The definition of the weak Pareto optimality [7] can be used to observe the optimality property.[2] Suppose that $u(\mu) \approxeq U_{\text{InP}}$ represent the utility achieved by matching μ. In case of no other matching μ' possible to meet the better utility, a matching μ is considered to be weak Pareto optimal, i.e., $u(\mu') \geq u(\mu)$.

Theorem 7.3 *Algorithm 2 produces a weak Pareto optimal (PO) solution for the problem presented in* (7.23).

Proof The proof is provided in Appendix E of the supplementary material of [6].

Computation Complexity and Practical Implementation

If the standard algorithm is applied for sorting of the preference profile for every MVNO user m_i, the complexity of constructing the preference profile is $\mathcal{O}(|\mathcal{K}| \log(|\mathcal{K}|))$ and in the same way, the complexity of building the preference profile at BS k for all MVNO users is $\mathcal{O}(|\mathcal{U}| \log(|\mathcal{U}|))$. So, the input to Algorithm 2 is $\xi = \sum_{k \in \mathcal{K}} |\mathcal{P}_k| + \sum_{m_i \in \mathcal{U}} |\mathcal{P}_{m_i}| = 2|\mathcal{K}||\mathcal{U}|$, where $|\mathcal{P}|$ represents the length of preference profile $\mathcal{P}$. As Algorithm 2 is finished on the completion of a given number of iterations (as given by the property of convergence in Property 1), it is noted that in the worst case, the time complexity of Algorithm 1 in Chap. 4 is quadratic (i.e., $\mathcal{O}(\xi) = \mathcal{O}(|\mathcal{K}||\mathcal{U}|) = \mathcal{O}(\max(|\mathcal{K}|, |\mathcal{U}|)^2)$), which is practically tolerable.

7.3.5 *Sub-channel Assignment and Energy Efficiency Optimization*

This section describes the analysis of the in-slice assignment of the sub-channel firstly and then the optimization problem for the energy efficiency of the given BS assignment solution. After that the development of an algorithm is explained in order to implement in a distributed way at each BS to allocate the resources (i.e., sub-channels) to its associated user so that the design objectives can be met, i.e., agreement of services and energy efficiency for the InP.

For that purpose the sub-channel assignment problem and the energy efficiency optimization problem can be formulated again for the given pricing decision and BS assignment solution, i.e., $\{\mathbf{p}^*, \mathbf{A}^*\}$, as follows:

[2]The optimality property in the proposed matching game only holds for the proposing side, i.e., BSs.

$$\max_{(\mathbf{Y},\mathbf{P})} V \frac{\sum\limits_{k\in\mathcal{K}} \sum\limits_{m_i\in\mathcal{S}_k} \sum\limits_{c\in\mathcal{C}} y_{m,i}^{(c)} R_{k,m,i}^{(c)}(\mathbf{P}^{(c)})}{\sum\limits_{k\in\mathcal{K}} (\sum\limits_{m_i\in\mathcal{S}_k} \sum\limits_{c\in\mathcal{C}} y_{m,i}^{(c)} P_k^{(c)} + P_k^0)} + \sum_{k\in\mathcal{K}} \sum_{m_i\in\mathcal{S}_k} \sum_{c\in\mathcal{C}} Q_{m,i} y_{m,i}^{(c)} R_{k,m,i}^{(c)}(\mathbf{P}^{(c)})$$

$$\text{s.t. } C_2, C_3, \tag{7.26}$$

$$C_5'' : \sum_{m_i\in\mathcal{S}_k} \sum_{c\in\mathcal{C}} y_{m,i}^{(c)} R_{k,m,i}^{(c)}(\mathbf{P}^{(c)}) \leq B_k, \ \forall k \in \mathcal{K},$$

$$C_6'' : \sum_{m_i\in\mathcal{S}_k} \sum_{c\in\mathcal{C}} y_{m,i}^{(c)} P_k^{(c)} \leq P_k^{\max}, \ \forall k \in \mathcal{K},$$

$$C_7'' : y_{m,i}^{(c)} \in \{0, 1\}, \ P_k^{(c)} \geq 0, \ \forall m, c, k.$$

It can be seen that the objective function in (7.26) is the fraction of two functions, which can be proven to be a non-convex function. It is noted that no standard scheme can be used to solve the non-convex optimization problems. To address this problem of efficient RA algorithm for the defined problem, a transformation from the given problem is proposed as follows.

Transformation of the Objective Function

Problem (7.26) is defined as a nonlinear fractional algorithm [3]. $\mathcal{F}$ is expressed as the possible set of solutions of problem (7.26) and expresses the maximum energy efficiency q^* of the selected InP as

$$q^* = \frac{\mathcal{R}(\mathbf{Y}^*, \mathbf{P}^*)}{\mathcal{P}(\mathbf{Y}^*, \mathbf{P}^*)} = \max_{(\mathbf{Y},\mathbf{P})} \frac{\mathcal{R}(\mathbf{Y}, \mathbf{P})}{\mathcal{P}(\mathbf{Y}, \mathbf{P})}, \ \forall \{\mathbf{Y}, \mathbf{P}\} \in \mathcal{F}. \tag{7.27}$$

Theorem 7.4 (Problem Equivalence) *The maximum energy efficiency q^* can be achieved when*

$$\max_{(\boldsymbol{Y},\boldsymbol{P})} \mathcal{R}(\boldsymbol{Y}, \boldsymbol{P}) - q^*\mathcal{P}(\boldsymbol{Y}, \boldsymbol{P}) = \mathcal{R}(\boldsymbol{Y}^*, \boldsymbol{P}^*) - q^*\mathcal{P}(\boldsymbol{Y}^*, \boldsymbol{P}^*) = 0, \tag{7.28}$$

for $\mathcal{R}(\boldsymbol{Y}, \boldsymbol{P}) \geq 0$ and $\mathcal{P}(\boldsymbol{Y}, \boldsymbol{P}) > 0$.

Proof Please refer to [3] for a proof of Theorem 7.4.

From Theorem 7.4, in a given problem having the objective function in the state of fraction, a replaceable objective function can be obtained in the state of subtraction, e.g., $\mathcal{R}(\mathbf{Y}, \mathbf{P}) - q^*\mathcal{P}(\mathbf{Y}, \mathbf{P})$. Therefore, the replaceable objective function can be considered to get the solution of the subproblem of sub-channel allocation and power assignment for MVNOs users.

Algorithm 3 Energy efficiency maximization algorithm

1: Initialization: $T_{\max}$, ϵ, $\tau = 0$, $q[0] = 0$, and Flag = **false**;
2: **repeat**
3: Solve problem (7.29) for a given q and obtain $\{\mathbf{Y}[\tau], \mathbf{P}[\tau]\}$;
4: **if** $\mathcal{R}(\mathbf{Y}[\tau], \mathbf{P}[\tau]) - q[\tau]\mathcal{P}(\mathbf{Y}[\tau], \mathbf{P}[\tau]) < \epsilon$ **then**
5: Flag = **true**;
6: **return** $\{\mathbf{Y}^*, \mathbf{P}^*\} = \{\mathbf{Y}[\tau], \mathbf{P}[\tau]\}$
7: $q^* = \frac{\mathcal{R}(\mathbf{Y}[\tau], \mathbf{P}[\tau])}{\mathcal{P}(\mathbf{Y}[\tau], \mathbf{P}[\tau])}$;
8: **else**
9: $q[\tau + 1] = \frac{\mathcal{R}(\mathbf{Y}[\tau], \mathbf{P}[\tau])}{\mathcal{P}(\mathbf{Y}[\tau], \mathbf{P}[\tau])}$ and $\tau = \tau + 1$;
10: Flag = **false**;
11: **end if**
12: **until** Flag = **true** or $\tau = T_{\max}$;

Iterative Algorithm for Energy Efficiency Maximization

This section presents an iterative method to solve the problem in (7.26) by using a known algorithm the Dinkelbach method [3]. For that purpose an equivalent objective function is used. The summary of the proposed algorithm given in Algorithm 3 and the guarantee of converging the algorithm to the optimal solution can be obtained by solving the inner problem (7.29) in each iteration. The proof of the convergence of Alg. 3 can be found in [3].

As given in Algorithm 3, on every step of the main loop, the following problem is solved with the value of parameter q:

$$\begin{aligned} &\max_{(\mathbf{Y},\mathbf{P})} \ \mathcal{R}'(\mathbf{Y}, \mathbf{P}) - q\mathcal{P}(\mathbf{Y}, \mathbf{P}) \\ &\text{s.t.} \quad C_2, C_3, C_5'', C_6'', C_7'', \end{aligned} \tag{7.29}$$

where

$$\mathcal{R}'(\mathbf{Y}, \mathbf{P}) = \sum_{k \in \mathcal{K}} \sum_{m_i \in \mathcal{S}_k} \sum_{c \in \mathcal{C}} (V + Q_{m,i}) y_{m,i}^{(c)} R_{k,m,i}^{(c)}(\mathbf{P}^{(c)}).$$

The given translated problem (7.29) is a mixed combinatorial and non-convex optimization problem where the non-convexity is due to the power allocation variables. These power variables are coupled by the cross-tier and co-tier interference given in the denominator of the SINR expression. Conversely, the integer used in the sub-channel assignment constraint causes the problem to be combinatorial in nature. In order to solve (7.29) the process is summarized in Algorithm 4.

The process is initiated by setting the values of the transmit power of BS k on all sub-channels c equally, i.e., $P_k^{(c)}[0] = P_k^{\max}/C$. By fixing the values of $P_k^{(c)}[t]$, problem (7.29) takes the form of a standard integer linear program which is possible to solve through a standard solver (line 3). It is assumed that there is a sub-channel assignment policy in order to find $\mathbf{Y}[t]$; problem (7.29) is still a non-convex problem

Algorithm 4 Procedure to solve (7.29)

1: Initialize $P_k^{(c)}[0] = P_k^{\max}/C$, and iteration $t = 0$;
2: **repeat**
3: Solve problem (7.29) for a fixed $P_k^{(c)}[t]$ and obtain $\mathbf{Y}[t]$;
4: Update $P_k^{(c)}[t+1]$ by solving (7.29) for given $\mathbf{Y}[t]$;
5: Set $t = t + 1$;
6: **until** Convergence of $\mathbf{P}$ and $\mathbf{Y}$;

and difficult to solve directly (line 4). Therefore DC programming-based solution is proposed in order to solve the non-convex power allocation problem [8, 12].

Theorem 7.5 *For a feasible problem (7.29), Algorithm 4 will converge to a local maximum of (7.29).*

Proof Please see a similar proof in [12].

D.C. Based Power Allocation

The non-concavity of the rate function in (7.5) for the variable $\mathbf{P}$ makes problem (7.29) non-convex for a fixed policy $\mathbf{Y}[t]$ of sub-channel allocation. To address this issue, a successive convex approximation technique [8, 12] is applied in order to compute the optimal power allocation in a separate time scale t_p.

At first, rate function (7.5) is expressed in D.C. form as

$$\sum_{c\in\mathcal{C}} y_{m,i}^{(c)} R_{k,m,i}^{(c)}(\mathbf{P}^{(c)}) = f_k(\mathbf{P}) - g_k(\mathbf{P}), \tag{7.30}$$

where $f_k(\mathbf{P})$ and $g_k(\mathbf{P})$ are the two concave functions defined as follows:

$$f_k(\mathbf{P}) = \sum_{c\in\mathcal{C}} y_{m,i}^{(c)} \ln\left(\sigma_{m,i}^{(c)} + \sum_{j\in\mathcal{K}} h_{j,m,i}^{(c)} P_j^{(c)}\right), \tag{7.31}$$

$$g_k(\mathbf{P}) = \sum_{c\in\mathcal{C}} y_{m,i}^{(c)} \ln\left(\sigma_{m,i}^{(c)} + \sum_{j\in\mathcal{K}, j\neq k} h_{j,m,i}^{(c)} P_j^{(c)}\right). \tag{7.32}$$

After that, the approximation is employed [8]

$$g_k(\mathbf{P}) \approx g_k(\mathbf{P}[t_p-1]) + \nabla g_k^T(\mathbf{P}[t_p-1])(\mathbf{P} - \mathbf{P}[t_p-1]) \tag{7.33}$$

for the power $\mathbf{P}[t_p-1]$ from iteration $t_p - 1 \geq 0$. Here, the vector with the length $(K+1)C$ represents the gradient $\nabla g_k^T(\mathbf{P})$ having elements defined as

$$\nabla g_k(\mathbf{P})^{(Cj+c)} \triangleq \begin{cases} 0, & \text{if } j = k, \\ \frac{y_{m,i}^{(c)} h_{j,m,i}^{(c)}}{\sum_{l \in \mathcal{K} \setminus \{k\}} h_{l,m,i}^{(c)} P_l^{(c)} + \sigma_{m,i}^{(c)}}, & \text{if } j \in \mathcal{K} \setminus \{k\}, \end{cases} \tag{7.34}$$

for $c \in \mathcal{C}$. From (7.30) and (7.33), we have

$$\begin{aligned} \sum_{c \in \mathcal{C}} R_{k,m,i}^{(c)}(\mathbf{P}^{(c)}) \approx\ & f_k(\mathbf{P}) - g_k(\mathbf{P}[t_p - 1]) \\ & - \nabla g_k^T(\mathbf{P}[t_p - 1])(\mathbf{P} - \mathbf{P}[t_p - 1]). \end{aligned} \tag{7.35}$$

The right-hand side of the above equation shows a concave function with respect to $\mathbf{P}$.

The approximate expression given in (7.35) helps to transform the non-convex problem (7.29) to a set of convex optimization subproblems with respect to $\mathbf{P}$. To solve for $\mathbf{P}$, the optimal solution $\mathbf{P}[t_p]$ is obtained at iteration $t_p > 0$ after starting from a feasible $\mathbf{P}[0]$ using the following convex program:

$$\begin{aligned} \max_{\mathbf{P}}\ & \sum_{k \in \mathcal{K}} \sum_{m_i \in \mathcal{S}_k} \sum_{c \in \mathcal{C}} (V + Q_{m,i}) R_{k,m,i}^{(c)}(\mathbf{P}^{(c)}) \\ & - q \sum_{k \in \mathcal{K}} \sum_{m_i \in \mathcal{S}_k} \sum_{c \in \mathcal{C}} \left(y_{m,i}^{(c)} P_k^{(c)} + P_k^0 \right) \\ \text{s.t.}\ & \sum_{m_i \in \mathcal{S}_k} \sum_{c \in \mathcal{C}} R_{k,m,i}^{(c)}(\mathbf{P}^{(c)}) \le B_k,\ \forall k \in \mathcal{K}, \\ & \sum_{m_i \in \mathcal{S}_k} \sum_{c \in \mathcal{C}} y_{m,i}^{(c)} P_k^{(c)} \le P_k^{\max},\ \forall k \in \mathcal{K}, \\ & 0 \ge P_k^{(c)} \ge 0,\ \forall k \in \mathcal{K}, c \in \mathcal{C}. \end{aligned} \tag{7.36}$$

$$P_k^{(c)}[t_s+1] = \left[\frac{\sum_{m_i \in \mathcal{S}_k} (V + Q_{m,i})(1 + \mu_k[t_s]) y_{m,i}^{(c)} h_{k,m,i}^{(c)} P_k^{(c)}[t_s] \left(\frac{1}{\mathcal{Q}_{k,m,i}^{(c)}(\mathbf{P}^{(c)}[t_s])} \right)}{(q + \nu_k[t_s]) \sum_{m_i \in \mathcal{S}_k} y_{m,i}^{(c)} + \sum_{j \in \mathcal{K} \setminus \{k\}} \sum_{n_i \in \mathcal{S}_j} (V + Q_{n,i})(1 + \mu_j[t_s]) y_{n,i}^{(c)} h_{k,n,i}^{(c)} \left(\frac{1}{\mathcal{I}_{j,n,i}^{(c)}(\mathbf{P}_{-j}^{(c)}[t_p - 1])} \right)} \right]_0^{P_k^{\max}} \tag{7.37}$$

$$\mu_k[t_s+1] = \left[\mu_k[t_s] + \kappa_\mu \left\{ \sum_{m_i \in \mathcal{S}_k} \sum_{c \in \mathcal{C}} y_{m,i}^{(c)} \ln \left(\frac{\mathcal{Q}_{k,m,i}^{(c)}(\mathbf{P}^{(c)}[t_s])}{\mathcal{I}_{k,m,i}^{(c)}(\mathbf{P}_{-k}^{(c)})[t_p - 1]} \right) - \frac{y_{m,i}^{(c)} \mathcal{I}_{k,m,i}^{(c)}(\mathbf{P}_{-k}^{(c)}[t_s])}{\mathcal{I}_{k,m,i}^{(c)}(\mathbf{P}_{-k}^{(c)})[t_p - 1]} + |\mathcal{C}| - B_k \right\} \right]^+ \tag{7.38}$$

$$\nu_k[t_s + 1] = \left[\nu_k[t_s] + \kappa_\nu \left(\sum_{m_i \in \mathcal{S}_k} \sum_{c \in \mathcal{C}} y_{m,i}^{(c)} P_k^{(c)}[t_s] - P_k^{\max} \right) \right]^+ \tag{7.39}$$

Algorithm 5 Distributed power allocation at BS $k \in \mathcal{K}$

1: Initialize: $\{\mu_k[0], \nu_k[0]\} > 0, \forall c \in \mathcal{C}, \forall m_i \in \mathcal{S}_k, P_k^{(c)}[0] = 0, \tilde{P}_k^{(c)}[0] = 0, t_p = 0$ and $t_s = 0$;
2: **repeat**
3: **repeat**
4: User $m_i(c) \in \mathcal{S}_k$ measures and reports $h_{k,m,i}^{(c)}$ and $\mathcal{I}_{k,m,i}^{(c)}(\tilde{\mathbf{P}}_{-k}^{(c)}[t_s])$ to BS $k \in \mathcal{K}$;
5: Compute $\mathcal{Q}_{k,m,i}^{(c)}(\tilde{\mathbf{P}}^{(c)}[t_s])\ \forall c \in \mathcal{C}$ and broadcasts to other BSs $j \in \mathcal{K}\backslash\{k\}$;
6: Compute $\tilde{P}_k^{(c)}$ by (7.37) $\forall c \in \mathcal{C}$;
7: Update μ_k, ν_k according to (7.38) and (7.39), respectively;
8: Set $t_s := t_s + 1$;
9: **until** μ_k, ν_k converge
10: Set $\mathbf{P}_{(k)}[t_p] = \tilde{\mathbf{P}}_{(k)}[t_s]$;
11: Broadcast $\mathcal{I}_{k,m,i}^{(c)}(\mathbf{P}_{-k}^{(c)}[t_p])$ to other BSs $j \in \mathcal{K}\backslash\{k\}$;
12: Set $t_p := t_p + 1$;
13: **until** $\mathbf{P}_{(k)}$ converge;

To get the solution, the optimal value of the transmit power of BSs and the Lagrange multipliers are obtained by using the KKT conditions and duality method [1] as given in (7.37), (7.38), and (7.39), respectively.

From the given expressions, a distributed power allocation scheme is proposed as shown in Algorithm 5.

Remark 7.3

1. Interference terms $\mathcal{I}_{k,m,i}^{(c)}(\mathbf{P}_{-k}^{(c)})$ locally measured and fed back by UE $m_i \in \mathcal{S}_k$, where we define

$$\mathcal{I}_{k,m,i}^{(c)}(\mathbf{P}_{-k}^{(c)}) \triangleq \sum_{l \in \mathcal{K}\backslash\{k\}} h_{l,m,i}^{(c)} P_l^{(c)} + \sigma_{m,i}^{(c)}. \tag{7.40}$$

2. Aggregate interference terms $\mathcal{Q}_{k,m,i}^{(c)}(\mathbf{P}^{(c)})$ broadcast by BS $k \in \mathcal{K}$, where we define

$$\mathcal{Q}_{k,m,i}^{(c)}(\mathbf{P}^{(c)}) \triangleq \mathcal{I}_{k,m,i}^{(c)}(\mathbf{P}_{-k}^{(c)}) + h_{k,m,i}^{(c)} P_k^{(c)}. \tag{7.41}$$

3. Channel gains $h_{k,m,i}^{(c)}$ measured and fed back by UE $m_i \in \mathcal{S}_k$. A block fading model is assumed where the channel information is required to send one time only at the start of Algorithm 5 [12].

Theorem 7.6 *For a given sub-channel assignment policy* $\mathbf{Y}[t]$*, Algorithm 5 converges to a locally optimal solution* $\mathbf{P}[t]$ *of problem (7.29).*

Proof The proof is provided in Appendix F of the supplementary material of [6].

7.3.6 Price Decision Optimization

The mean time throughput of the MVNOs users (or the expected value of the throughput from serving the MVNOs users) is used to compute the price offered to MVNOs in an independent MVNO model. Therefore, the optimization problem for short-term decisioning of the price with the constraints on the stability of the contract agreement queue is formulated as follows:

$$\max_{\beta(t))\in\Omega} \sum_{m\in\mathcal{M}} \sum_{i\in\mathcal{U}_m} \left[V\beta_m(t) r_{m,i}(t) - Q_{m,i}(t) r_{m,i}(t) \right], \tag{7.42}$$

where $\Omega = \{\beta_m, m \in \mathcal{M} | \beta^{\min} \leq \beta_m \leq \beta^{\max}\}$, and $r_{m,i}(t)$ from (7.9). The former expression in the objective function refers to the profit of the InP obtained from serving MVNOs users, and the latter term denotes the cost of the contract agreement queue backlog. InP profit is optimized by maximizing the given objective function while pushing the virtual queue backlog in such a way to satisfy the stability constraint and assure the MVNOs agreement of services. It is to be noted that when the competing MVNOs model is considered where the users' rates are computed in Sect. 3.2, it is not required to solve the optimization problem for the price decisioning. The mean time throughput of user i of MVNO m for each pricing period is given as

$$\mathbb{E}(R_{m,i}(t)) = \int_0^{R_{m,i}^{\max}} R_{m,i} f(R_{m,i}) dR_{m,i}, \tag{7.43}$$

where $R_{m,i}^{\max}$ gives the upper bound of the transmission rate of each user i of MVNO m till time t, respectively, and $f(R_{m,i})$ represents the probability density function of $R_{m,i}$. The transmission rate is the function of SINR, which is dependent on the distributions of all power channel gains (is exponentially distributed because of the amplitude of channel gain following a Rayleigh distribution).

The operator $[\cdot]_x$ of $r_{m,i}(t)$ in (7.9) makes the problem (7.42) non-convex. For convenience, a new variable is defined

$$z_m(t) = \begin{cases} 1, \text{ if } \beta_m(t) < \dfrac{\mathfrak{B}_m^{\max}}{|\mathcal{U}_m| r_{m,i}^{\min}}, \\ 0, \quad\quad \text{otherwise.} \end{cases} \tag{7.44}$$

After that, $r_{m,i}(t) = \mathfrak{B}_m^{\max} / |\mathcal{U}_m| \beta_m(t)$ when $z_m(t) = 1$, and $r_{m,i}(t) = r_{m,i}^{\min}$ when $z_m(t) = 0$. Without loss of generality, it is assumed that all users connected to the same MVNO are meeting the homogeneous minimum data rate demand, i.e., $r_{m,i}^{\min} = r_m^{\min}, \forall i \in \mathcal{U}_m$. Therefore, the profit of the InP at time t is given as

$$\mathcal{R}_m(\beta_m(t), z_m(t)) = z_m(t) \frac{\mathfrak{B}_m^{\max}}{|\mathcal{U}_m|} + (1 - z_m(t)) \beta_m(t) r_m^{\min}. \tag{7.45}$$

Algorithm 6 Price decision algorithm

```
1: initialize: t = 0; Q = {Q_init}; L_0 = Φ_lb(Q_init); U_0 = Φ_ub(Q_init); ε > 0;
2: repeat
3:     t ← t + 1;
4:     Q_t = {Q ∈ Q | m = argmin(|z*_m − 1/2|)};
5:     Q_t^(0) = {β, z | z_m = 0}; Q_t^(1) = {β, z | z_m = 1};
6:     Q = {Q \ Q_t} ∪ {Q_t^(0), Q_t^(1)};
7:     for Q_t^(i), i ∈ {0, 1} do
8:         Calculate Φ_lb(Q_t^(i)) and Φ_ub(Q_t^(i));
9:     end for
10:    U_t = min(U_t, Φ_ub(Q_t^(i)), i = 0, 1);
11:    L_t = min(L_t, Φ_lb(Q_t^(i)), i = 0, 1);
12:    Q^(pru) = {Q ∈ Q | Φ_lb(Q) ≥ U_t)};
13:    Q = {Q \ Q^(pru)};
14: until U_t − L_t ≤ ε;
```

The offered price computed at time t can be calculated by finding the solution of the following problem:

$$\max_{(\boldsymbol{\beta},\mathbf{z})} \sum_{m\in\mathcal{M}} \left[V\mathcal{R}_m(\beta_m(t), z_m(t)) - \sum_{i\in\mathcal{U}_m} \mathcal{Q}_{m,i}(t) r_{m,i}(t) \right], \tag{7.46}$$

where V is a controlled parameter.

It can be seen that problem (7.46) is a mixed-boolean program, which means that it needs exponential computation services in order to get the optimal solution by using the exhaustive search. This motivation allows to propose an efficient scheme in Algorithm 6 to get the solution of this problem.

BnB method [1] is used to develop Algorithm 6, which allows to compute the global optimal solution for the problem of price decisioning (7.42). Suppose $\mathcal{Q}_{init} = \{\boldsymbol{\beta}, \mathbf{z}\}$ be the original search space, containing all of the possible combinations of indicator variable $\mathbf{z}$. A set of subdomains $\boldsymbol{Q} = \{\mathcal{Q}_t \subset \mathcal{Q}_{init}, t = 1, 2, \ldots\}$ are maintained in the proposed algorithm, where t denotes the iteration of the algorithm. For every $\mathcal{Q}_t$, suppose $\Phi_{ub}(\cdot)$ and $\Phi_{lb}(\cdot)$ represent the upper and lower bounds. $\Phi_{ub}(\mathcal{Q}_t)$ and $\Phi_{lb}(\mathcal{Q}_t)$ are referred as the local upper and lower bounds, respectively, which are corresponding to subdomain $\mathcal{Q}_t$.

The algorithm is initiated and boolean variable z_m is relaxed, i.e., $0 \leq z_m \leq 1, \forall m \in \mathcal{M}$, and after that the relaxed problem is solved in order to obtain a lower bound $L_0 = \Phi_{lb}(\mathcal{Q}_{init})$ for the actual problem (7.46). After that, optimal relaxed variables z_m^* are rounded off to 0 or 1, $\forall m \in \mathcal{M}$, and (7.46) is solved again by using these fixed values of z_m^* to get the upper bound $U_0 = \Phi_{ub}(\mathcal{Q}_{init})$ (line 2). On every iteration t, the search space is split into two subspaces $\mathcal{Q}_t^{(0)}$ and $\mathcal{Q}_t^{(1)}$ by choosing $\mathcal{Q}_t \in \boldsymbol{Q}$ in a way to get $m = \text{argmin}(|z_m^* - 1/2|)$, then modified $\boldsymbol{Q}$ by removing $\mathcal{Q}_t$ (lines 5–7). The lower and upper bounds are calculated for each subspace and the one with the smallest lower bound is selected (lines 8–12). At the end, the subspaces satisfying $\Phi_{lb}(\mathcal{Q}) \geq U_t$ are rejected, because every point in these spaces leads to a degraded performance as compared to the current upper bound (lines 13–14). The algorithm is terminated if $U_t - L_t \leq \epsilon$.

Lemma 7.3 *Algorithm 6 converges to the optimal solution of price decision problem (7.42).*

Proof The proof is similar to the one provided in [1].

7.4 Simulation Results

7.4.1 Simulation Setting

This section presents the performance evaluation of the proposed scheme through simulations. Figure 7.2 explains the placement of MVNOs subscribed users which are randomly placed in circle of a macro cell having radius of 500 m and it is covering four SBSs where all SBSs have radii of 100 m. All of these macro and small-cell base stations (BSs) are owned by the infrastructure provider. It is considered that multiple MVNOs each having a varied number of users. Each SBS and the MBS have a distance of 250 m between them, and an SBS at least $250\sqrt{2}$ m

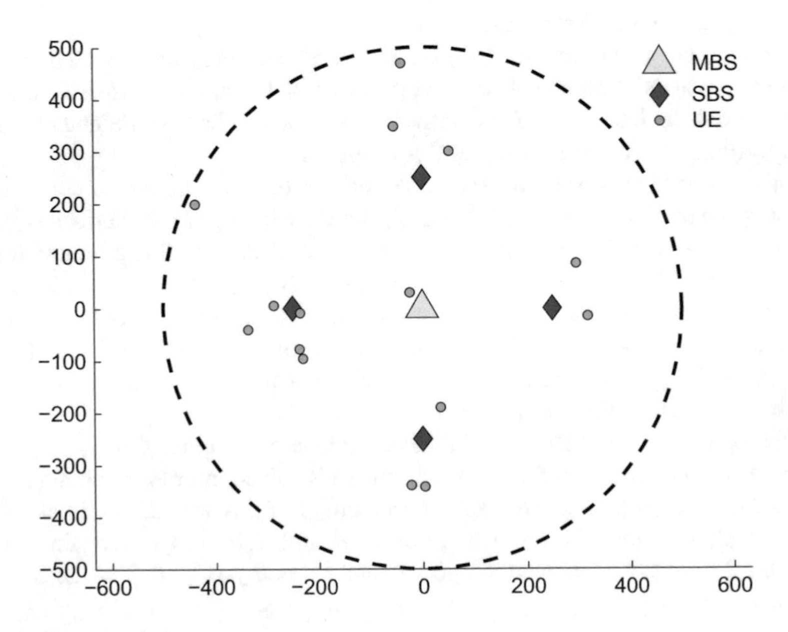

Fig. 7.2 Network topology

away from the other SBS. It is assumed that the InP has $|\mathcal{C}| = 10$ OFDMA sub-channels, each having a total bandwidth of 180 KHz. Power is considered $P_0^{\max} = 40$ W and $P_k^{\max} = 4$ W, $\forall m \in \mathcal{M}\backslash\{0\}$. Every sub-channel $c \in \mathcal{C}$ has a noise power of 10^{-13} W. The links of small-scale fading coefficients from the BS-to-user are i.i.d Rayleigh distributed with variance equal to unity. Channel gains value is given as $h_{k,m,i}^{(c)} = \chi^{(c)} d_{k,m,i}^{-\alpha}$, where $\chi^{(c)}$ is used to create the random value, $d_{k,m,i}$ represents the physical distance among BS k and user i of MVNO m, and $\alpha = 3$ is giving the path-loss exponent.

To model the backhaul network, commercial optical fiber modem specifications [11] are used where three different levels of data rates are supported for backhaul of distance less than 2.5 km: $R_1 = 11.184$ Mbps, $R_2 = 34.368$ Mbps, and $R_3 = 44.736$ Mbps. The fixed backhaul capacity of MBS is set to $B_0 = R_3$; however, the backhaul capacities for the SBSs are varied between R_1, R_2, and R_3 for the performance analysis of the proposed framework using various backhaul capacities. A star backhaul link topology is considered where all BSs are connected directly to the main controller through individual backhaul links. It is assumed that each MVNOs has a minimum data rate range $r_{\min} = 5, 8$, and 10 bps/Hz/user. The price is bounded between $\beta_{\min} = 0.1$ and $\beta_{\max} = 0.3$ and the scaling factor θ of the price is considered as 2.

The proposed energy efficiency framework is "Proposed EE" and it is evaluated by two baseline algorithms for comparison. The former baseline algorithm is the maximum power allocation (represented by "Maximum power"), i.e., each sub-channel is set an equal and fixed transmit power $P_k^{\max}/|\mathcal{S}_k|$, and the derived optimal power allocation in (7.37) is not utilized. The other baseline algorithm is maximizing the network sum-rate, i.e., the numerator of η_{EE} in (7.7) (denoted by "Sum-rate") with the same constraints in (7.26).

7.4.2 Numerical Results

In Fig. 7.3, it is seen that as the number of users is increased in the network, there is a decrease in the average throughput of each user. This is due to the fact that limited resources are shared among the increasing number of users causing more interference in the network. Furthermore, in Fig. 7.4, it is observed that, as the number of users is increased in the network, there is an increase in the average total throughput, until the saturation point is reached by making the number of MVNOs and/or the number of users large enough in the network up to 20 and above. The reason behind the throughput saturation is the limitations of network resources in the existing simulation setting. Moreover, it is observed that the proposed approach can get distinct throughput as compared to the throughput of baseline and when compared to the maximum power baseline approach, it can get a significant performance gain.

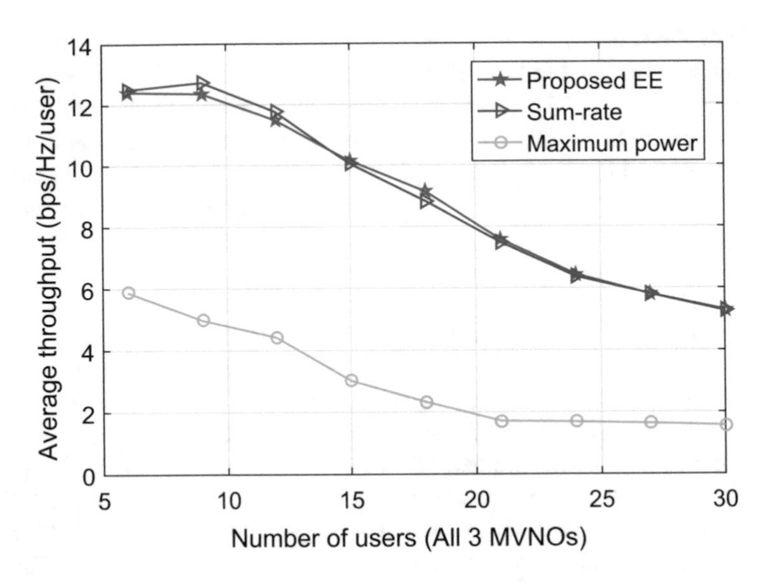

Fig. 7.3 Average throughput (bps/Hz/user)

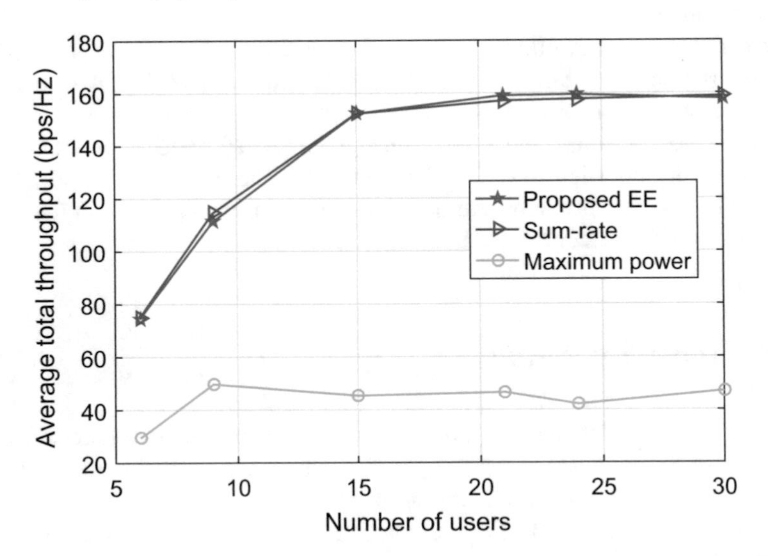

Fig. 7.4 Average total network throughput (bps/Hz)

In Fig. 7.5, the average energy efficiency is evaluated where the number of users in the network is increased. With the increase in the number of users in the network, the energy efficiency is also increased. This is due to the fact that the increase in the number of network users occupies the underutilized resources, therefore it also increases the energy efficiency. On the other hand, the utilization of all of the resources of the network causes the saturation of energy efficiency when the number

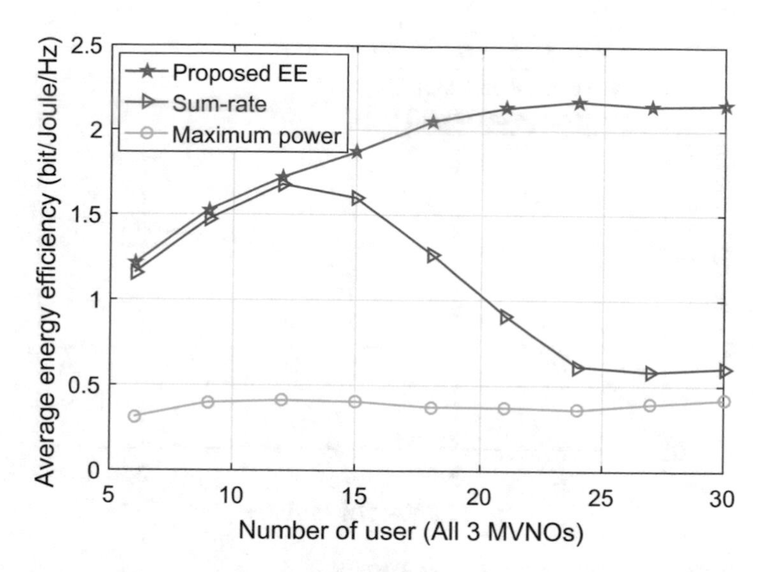

Fig. 7.5 Average energy efficiency (bit/Joule/Hz)

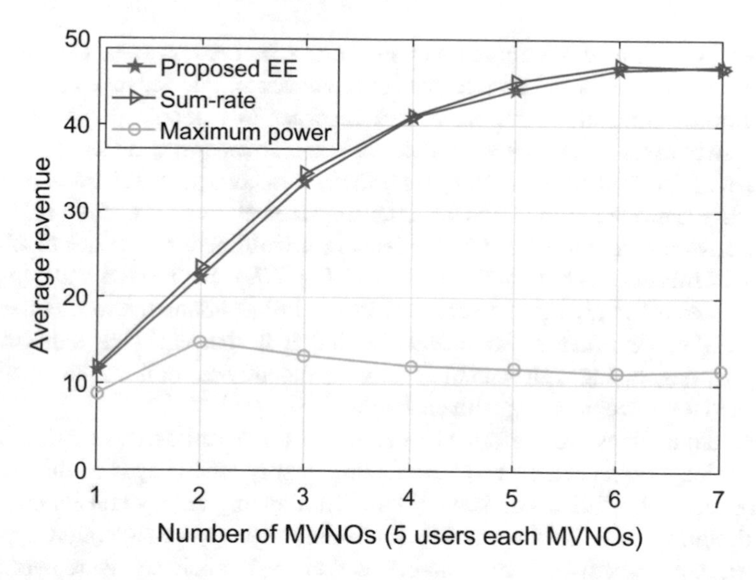

Fig. 7.6 Average revenue

of users in the network is 20 and above. In addition, both of the baselines face an increase in the power consumption and reduction in energy efficiency with the increase in the number of network users.

Moreover, in Fig. 7.6, the same trend is observed in terms of average profit gain, i.e., after reaching at the saturation point of total network throughput the profit is also saturated. As the revenue is dependent on the data rate, same revenue gains are

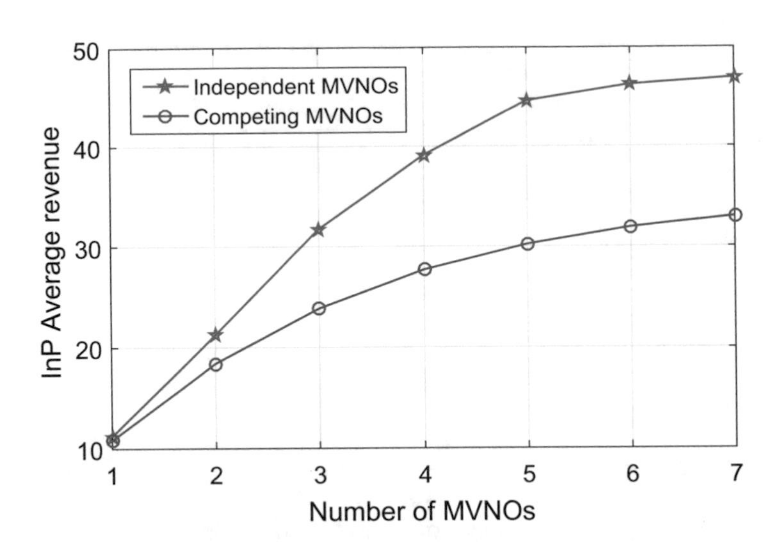

Fig. 7.7 Independent MVNOs vs competing MVNOs

obtained by both of the proposed and sum-rate scheme, however, low revenue is obtained because of the achievable rate for the maximum power scheme.

Additional simulations are performed in order to compare the InPs revenue for the two cases of MVNOs model: the first is independent MVNOs model (represented by "Independent MVNOs") and the second is competition as a non-cooperative game between MVNOs model (represented by "Competing MVNOs"). The parameters are normalized in the leasing price function in order to keep the revenue of InP same in both of the models. In Fig. 7.7, it can be seen that competing MVNOs scheme is giving more revenue to the InP as compared to no interaction between MVNOs scheme. The reason behind it is that each InP can get more revenue by negotiating with each MVNO independently as compared to the revenue when MVNOs are competing with each other.

Next, simulations are performed to compare the performances under different design objectives, represented by maximizing energy efficiency (i.e., the proposed model denoted by "EE maximization") and minimizing energy consumption. The same design is used for the baseline minimizing energy consumption approach; however, the objective of maximizing the energy efficiency is replaced with minimizing energy cost for all of the BSs, i.e., minimizing energy consumption (represented by "Energy minimization"). Furthermore, the results are compared with the sum-rate maximization baseline (represented by "Sum-rate maximization").

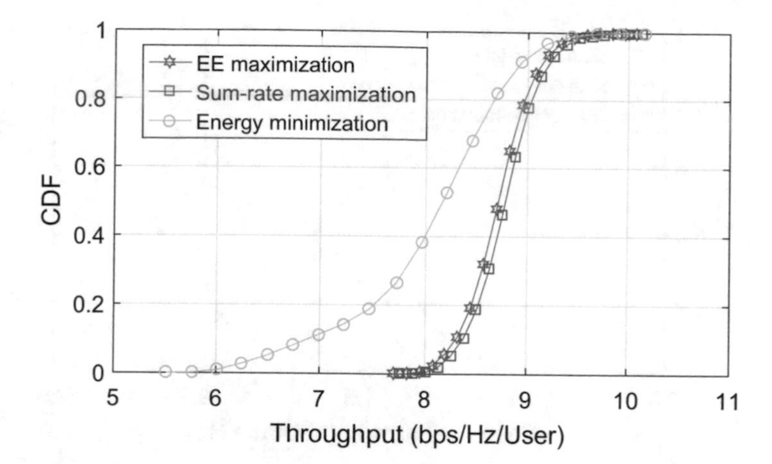

Fig. 7.8 Throughput

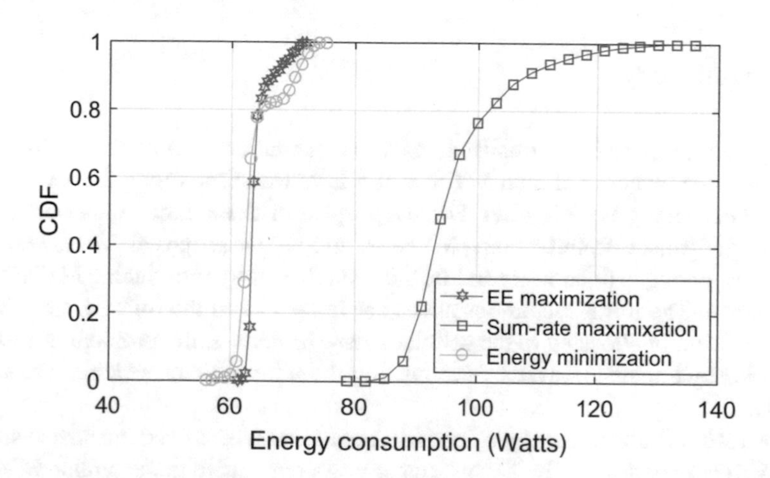

Fig. 7.9 Energy consumption

In Figs. 7.8 and 7.9, it is observed that the proposed "EE maximization" approach gets low energy consumption same as the baseline "Energy minimization" approach, while getting the same throughput as compared to sum-rate maximization baseline. Furthermore, there is significant gain in average throughput as compared to the baseline "Energy minimization" approach (Fig. 7.8). Therefore, the proposed "EE maximization" scheme is getting the best energy efficiency as compared to the baseline schemes while considering the high throughput gain and the consumption of energy, i.e., "Sum-rate maximization" and "Energy minimization," by 46% and 12%, respectively (Fig. 7.10).

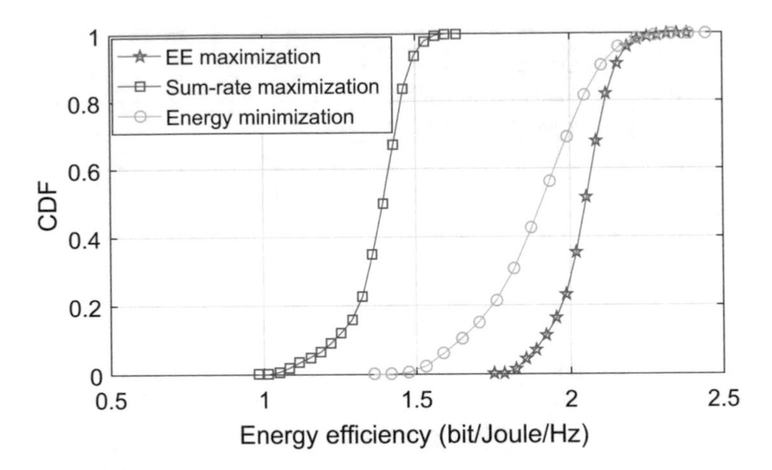

Fig. 7.10 Energy efficiency

7.5 Summary

This paper proposed a virtualized resource allocation (RA) scheme to jointly maximize the revenues of both MVNOs and InPs while satisfying the service level agreements among MVNOs/InPs. For that purpose, a hierarchical framework having three participants of cellular users, MVNOs, and InPs is proposed. Virtualization of resources among cellular users and InPs is established by introducing MVNOs as a middleman. The infrastructure resources are bought from the InPs by the MVNOs and are virtually allocated to the cellular users. Moreover, the problem of resource allocation by the InP is solved by formulating the sub-problems which can be solved locally.

The resource allocation algorithm is designed first where the demand decisioning of MVNOs is performed by formulating a non-cooperative game among MVNOs. After that, the price decisioning problem for the InPs is solved by decomposing the global problem into short-term local problems. Then matching theory is used for BS assignment to the MVNOs, and sub-channel/power allocation is performed using convex optimization.

References

1. Boyd, S., & Vandenberghe, L. (2004). *Convex optimization*. Cambridge: Cambridge University Press.
2. Damnjanovic, A., Montojo, J., Wei, Y., Ji, T., Luo, T., Vajapeyam, M., et al. (2011). A survey on 3GPP heterogeneous networks. *IEEE Wireless communications, 18*(3), 10–21.
3. Dinkelbach, W. (1967). On nonlinear fractional programming. *Management Science, 13*(7), 492–498.

4. Granelli, F., Gebremariam, A. A., Usman, M., Cugini, F., Stamati, V., Alitska, M., et al. (2015). Software defined and virtualized wireless access in future wireless networks: Scenarios and standards. *IEEE Communications Magazine, 53*(6), 26–34.
5. Gu, Y., Saad, W., Bennis, M., Debbah, M., & Han, Z. (2015). Matching theory for future wireless networks: Fundamentals and applications. *IEEE Communications Magazine, 53*(5), 52–59.
6. Ho, T. M., Tran, N. H., Le, L., Han, Z., Kazmi, S. A., & Hong, C. S. (2018). Network virtualization with energy efficiency optimization for wireless heterogeneous networks. *IEEE Transactions on Mobile Computing*. https://ieeexplore.ieee.org/document/8476209/authors#authors
7. Jorswieck, E. A. (2011). Stable matchings for resource allocation in wireless networks. In *2011 17th International Conference on Digital Signal Processing (DSP)* (pp. 1–8). Piscataway: IEEE.
8. Kha, H. H., Tuan, H. D., & Nguyen, H. H. (2012). Fast global optimal power allocation in wireless networks by local DC programming. *IEEE Transactions on Wireless Communications, 11*(2), 510–515.
9. Liang, C., & Yu, F. R. (2015). Wireless network virtualization: A survey, some research issues and challenges. *IEEE Communications Surveys & Tutorials, 17*(1), 358–380.
10. Neely, M. J. (2010). Stochastic network optimization with application to communication and queueing systems. *Synthesis Lectures on Communication Networks, 3*(1), 1–211.
11. Ng, D. W. K., Lo, E. S., & Schober, R. (2012). Energy-efficient resource allocation in OFDMA systems with large numbers of base station antennas. *IEEE Transactions on Wireless Communications, 11*(9), 3292–3304.
12. Ngo, D. T., Khakurel, S., & Le-Ngoc, T. (2014). Joint subchannel assignment and power allocation for OFDMA femtocell networks. *IEEE Transactions on Wireless Communications, 13*(1), 342–355.
13. Roth, A. E. (2008). Deferred acceptance algorithms: History, theory, practice, and open questions. *International Journal of Game Theory, 36*(3–4), 537–569.
14. Zhu, K., & Hossain, E. (2016). Virtualization of 5G cellular networks as a hierarchical combinatorial auction. *IEEE Transactions on Mobile Computing, 15*(10), 2640–2654.

Chapter 8
Concluding Remarks

In this chapter, we conclude our book by summarizing the key insights on network slicing and 5G networks. This book will provide a comprehensive guide to the emerging field of network slicing and its importance to bring novel 5G applications into fruition. It discusses the current trends, novel enabling technologies, and current challenges imposed on the cellular networks. Then, we discuss about the concept of network slicing, the enabling technologies which can be applied in 5G networks to meet the stringent requirements posed by the end user devices. Resource management aspects of network slicing are also discussed by summarizing the recent research works. Finally, we also present use cases of network slicing and applications for vertical industries.

8.1 Open Issues

8.1.1 Dynamic Slice Allocation

The surveyed works [1–6] present efficient resource management schemes by providing network slices to MVNOs users. However, all these works have overlooked one important aspect of dynamic network slicing in 5G networks. These works mainly assume that the user generated demand is fixed during the resource management process. Thus, they have provided solutions for these fixed demands without considering a dynamic environment. However, static environment does not reflect the practical implementation of the system. A practical system would have users arriving and leaving a system with different demands at different time slots. Thus, the goal is to device novel approaches that can adopt to such dynamic environments, i.e., new users entering a system and demands generated from the users at different time slots. This will result in providing a heterogeneous size slice

S. M. A. Kazmi et al., *Network Slicing for 5G and Beyond Networks*,
https://doi.org/10.1007/978-3-030-16170-5_8

to the same user based on its current demands at a specific time slot. Moreover, it will also be able to cope with dynamic users entering and leaving the network system.

8.1.2 Mobility Aware Network slicing

One of the main goals in 5G networks is to provide services to end user devices in case of high mobility. However, the current approaches for network slicing are not designed to handle mobility in the network. Indeed handling and orchestrating the radio access and core network will be very challenging in case of mobility which would require migration of services from one point to other points in the network. Moreover, a strong coordination among multiple cells would also be required to handle such cases. Moreover, inter-InP coordination might also be required to handle mobility as a single operator InP might not have enough resources in a specific area to support its mobile users. Most of the recent research works discussed in this book overlooked the inter-InP coordination issue. These implications can help in developing a mobility aware network slicing for mobile end users.

8.2 Conclusion

This book compiles the research works for resource management in the future networks. To present a full view of the resource management problem in 5G networks, we first introduce the requirements and enabling technologies of 5G networks. Moreover, we also provide a detailed overview of network slicing that can be adopted to fulfill the 5G deliverables. Then we discuss recent research works their motivation, issues, challenges, and solutions. Furthermore, we present some open issues for the future research and their potential. Research on network slicing is still in its early stage. This book will help researchers and engineers from both academia and industry to better understand the current works and contribute more to this area.

References

1. Kazmi, S. A., Tran, N. H., Ho, T. M., & Hong, C. S. (2018). Hierarchical matching game for service selection and resource purchasing in wireless network virtualization. *IEEE Communications Letters, 22*(1), 121–124.
2. Kazmi, S. A., & Hong, C. S. (2017). A matching game approach for resource allocation in wireless network virtualization. In *Proceedings of the 11th International Conference on Ubiquitous Information Management and Communication* (p. 113). New York: ACM.

3. Ho, T. M., Tran, N. H., Kazmi, S. A., Han, Z., & Hong, C. S. (2018). Wireless network virtualization with non-orthogonal multiple access. In *NOMS 2018-2018 IEEE/IFIP Network Operations and Management Symposium* (pp. 1–9). Piscataway: IEEE.
4. Vo, P. L., Nguyen, M. N. H., Le, T. A., & Tran, N. H. (2018). Slicing the edge: Resource allocation for RAN network slicing. *IEEE Wireless Communications Letters, 7*(6), 970–973.
5. LeAnh, T., Tran, N. H., Ngo, D. T., & Hong, C. S. (2017). Resource allocation for virtualized wireless networks with backhaul constraints. *IEEE Communications Letters, 21*(1), 148–151.
6. Ho, T. M., Tran, N. H., Le, L., Han, Z., Kazmi, S. A., & Hong, C. S. (2018). Network virtualization with energy efficiency optimization for wireless heterogeneous networks. *IEEE Transactions on Mobile Computing*. https://ieeexplore.ieee.org/document/8476209/authors#authors

Index

S. M. A. Kazmi et al., *Network Slicing for 5G and Beyond Networks*,
https://doi.org/10.1007/978-3-030-16170-5

F

G

H

I

J

L

M

Printed in the United States
By Bookmasters